Capillary Electrophoresis of Proteins

CHROMATOGRAPHIC SCIENCE SERIES

A Series of Monographs

Editor: JACK CAZES
Cherry Hill, New Jersey

1. Dynamics of Chromatography, *J. Calvin Giddings*
2. Gas Chromatographic Analysis of Drugs and Pesticides, *Benjamin J. Gudzinowicz*
3. Principles of Adsorption Chromatography: The Separation of Nonionic Organic Compounds, *Lloyd R. Snyder*
4. Multicomponent Chromatography: Theory of Interference, *Friedrich Helfferich and Gerhard Klein*
5. Quantitative Analysis by Gas Chromatography, *Josef Novák*
6. High-Speed Liquid Chromatography, *Peter M. Rajcsanyi and Elisabeth Rajcsanyi*
7. Fundamentals of Integrated GC-MS (in three parts), *Benjamin J. Gudzinowicz, Michael J. Gudzinowicz, and Horace F. Martin*
8. Liquid Chromatography of Polymers and Related Materials, *Jack Cazes*
9. GLC and HPLC Determination of Therapeutic Agents (in three parts), *Part 1 edited by Kiyoshi Tsuji and Walter Morozowich, Parts 2 and 3 edited by Kiyoshi Tsuji*
10. Biological/Biomedical Applications of Liquid Chromatography, *edited by Gerald L. Hawk*
11. Chromatography in Petroleum Analysis, *edited by Klaus H. Altgelt and T. H. Gouw*
12. Biological/Biomedical Applications of Liquid Chromatography II, *edited by Gerald L. Hawk*
13. Liquid Chromatography of Polymers and Related Materials II, *edited by Jack Cazes and Xavier Delamare*
14. Introduction to Analytical Gas Chromatography: History, Principles, and Practice, *John A. Perry*
15. Applications of Glass Capillary Gas Chromatography, *edited by Walter G. Jennings*
16. Steroid Analysis by HPLC: Recent Applications, *edited by Marie P. Kautsky*
17. Thin-Layer Chromatography: Techniques and Applications, *Bernard Fried and Joseph Sherma*
18. Biological/Biomedical Applications of Liquid Chromatography III, *edited by Gerald L. Hawk*
19. Liquid Chromatography of Polymers and Related Materials III, *edited by Jack Cazes*
20. Biological/Biomedical Applications of Liquid Chromatography, *edited by Gerald L. Hawk*
21. Chromatographic Separation and Extraction with Foamed Plastics and Rubbers, *G. J. Moody and J. D. R. Thomas*
22. Analytical Pyrolysis: A Comprehensive Guide, *William J. Irwin*

Capillary Electrophoresis of Proteins

Tim Wehr
Bay Bioanalytical Laboratory
Richmond, California

Roberto Rodríguez-Díaz
Dynavax Technologies
Berkeley, California

Mingde Zhu
Bio-Rad Laboratories
Hercules, California

MARCEL DEKKER, INC.   NEW YORK · BASEL · HONG KONG

Library of Congress Cataloging-in-Publication Data

Wehr, Tim.
  Capillary electrophoresis of proteins / Tim Wehr, Roberto Rodríguez-Díaz, Mingde Zhu.
    p.   cm. — (Chromatographic science; v. 80)
    Includes bibliographical references and index.
    ISBN 0-8247-0205-0 (alk. paper)
    1. Proteins—Analysis. 2. Capillary electrophoresis. I. Rodríguez-Díaz, Roberto.
  II. Zhu, Mingde. III. Title. IV. Series.
  QP551.W396 1998
  572'.6—dc21
                                                                          98-27751
                                                                              CIP

This book is printed on acid-free paper.

Headquarters
Marcel Dekker, Inc.
270 Madison Avenue, New York, NY 10016
tel: 212-696-9000; fax: 212-685-4540

Eastern Hemisphere Distribution
Marcel Dekker AG
Hutgasse 4, Postfach 812, CH-4001 Basel, Switzerland
tel: 44-61-261-8482; fax: 44-61-261-8896

World Wide Web
http://www.dekker.com

The publisher offers discounts on this book when ordered in bulk quantities. For more information, write to Special Sales/Professional Marketing at the headquarters address above.

Current printing (last digit)
10 9 8 7 6 5 4 3 2 1

PRINTED IN THE UNITED STATES OF AMERICA

Preface

This book evolved from a review of capillary electrophoresis (CE) of proteins that was published in *Advances in Chromatography* in 1996. We recognized that although a literature review was timely and needed in this growing field, there was no comprehensive practical guide for researchers interested in applying CE to protein analysis. Our own experience had shown us that simply reading published research papers was necessary but not sufficient to achieve successful results in the laboratory. To fill this gap in the published information on capillary electrophoresis and to aid practitioners with limited expertise in CE of proteins, we undertook preparation of this work.

Capillary Electrophoresis of Proteins is designed to be both a reference and a handbook. It is organized by separation mode (zone electrophoresis, isoelectric focusing, and sieving) with discussions on separation principles, method development, and optimization. Practical details of buffer preparation, capillary selection and handling, and troubleshooting problem separations are emphasized. For readers who are new to the area of capillary electrophoresis, introductory chapters on the basic principles of capillary electrophoresis and CE instrumentation are included. Where appropriate, step-by-step methods are provided. The literature review we published in 1996 has been updated in this book to provide a current overview of published applications with 386 references. The book is intended to serve as an introduction to protein CE and as a useful reference for scientists applying CE to protein analysis on a regular basis.

Much of the practical information in this work originates from our experience in the capillary electrophoresis group at Bio-Rad Laboratories over the last decade. In addition to the development of CE instrumentation and separation methods, we worked continually with scientists in the field to apply CE to real-life bioanalytical problems. This book represents a distillation of our experience, which we hope will be beneficial to practicing scientists in both academic and industry environments.

Tim Wehr
Roberto Rodríguez-Díaz
Mingde Zhu

Contents

Capillary Electrophoresis of Proteins

1

Introduction

Separation technology has been central to the elucidation of protein structure and function. Both chromatography and electrophoresis have been used for decades to isolate and characterize proteins and their components. Open column and low pressure chromatography were essential for the purification of sufficient amounts for protein for structural analysis, and the preparation of high-resolution ion exchange resins enabled development of the amino acid analyzer which paved the way for analysis of protein composition. The introduction of protein-compatible ion exchange and size exclusion supports allowed high-performance liquid chromatography to be used for obtaining highly pure proteins, while high-efficiency reversed phase columns proved invaluable for high-resolution peptide mapping and purification, and for identification of the PTH amino acids generated by automated Edman sequence analysis. Gel electrophoresis evolved in parallel with chromatography, providing an inexpensive method for separating complex protein mixtures. The development of the Laemmli system for separating SDS-protein complexes on polyacrylamide gels proved to be such a powerful technique that it is used on a daily basis in virtually every protein chemistry laboratory in the world. Isoelectric focusing provided an alternative separation technique for separation of proteins based on their isoelectric points, and the combination of IEF and SDS-PAGE by O'Farrell resulted in a two-dimensional separation technique which has the power to resolve thousands of proteins on a single 2-D gel. The development of blotting techniques to transfer separated proteins from the gel to a

1

suitable support for other analyses such as immunoassay or sequencing greatly simplified many experiments in protein chemistry.

As powerful as these protein separation techniques are, they are not without limitations. Chromatographic separations are based on interaction of an analyte with the surface of the stationary phase. However, proteins are by nature very surface-active molecules. They also possess low diffusion constants and display poor mass transfer kinetics during the chromatographic process. As a consequence, resolution is often less than desired, protein recovery may be low, and native proteins may be denatured during the separation. Gel electrophoresis, on the other hand, is a laborious and time-consuming technique requiring preparation of the gel, separation of the sample, staining and destaining, and gel-drying. The gel must be treated with a dye or stain to visualize the separated proteins and because the uptake of stain may occur in a nonlinear fashion, the intensity of the stained bands may be poorly correlated with amount of protein. For this reason, gel electrophoresis is, at best, a semiquantitative technique.

Capillary electrophoresis (CE) is a relatively new separation technology which combines aspects of both gel electrophoresis and HPLC. Like gel electrophoresis, the separation depends upon differential migration in an electrical fied. Since its first description in the late 1960s, capillary electrophoretic techniques analogous to most conventional electrophoretic techniques have been demonstrated: zone electrophoresis, displacement electrophoresis, isoelectric focusing, and sieving separations. Unlike conventional electrophoresis, however, the separations are performed in free solution without the requirement for a casting a gel. As in HPLC, detection is accomplished as the separation progresses, with resolved zones producing an electronic signal as they migrate past the monitor point of a concentration-sensitive (e.g., UV absorbance or fluorescence) detector. Therefore the need for staining and destaining is eliminated. Data presentation and interpretation is therefore also similar to HPLC; the output (peaks on a baseline) can be displayed as an electropherogram and integrated to produce quantitative information in the form of peak area or height. In CE and HPLC, a single sample is injected at the inlet of the capillary and multiple samples are analyzed in serial fashion. This contrasts to conventional electrophoresis in which multiple samples are frequently run in parallel as lanes on the same gel. This limitation of CE in sample throughput is compensated by the ability to process samples automatically using an autosampler. Compared to its elder cousins, CE is characterized by high resolving power, sometimes higher than electrophoresis or HPLC. The use of narrow-bore capillaries with excellent heat dissipation properties enable the use of very high field strengths (sometimes in excess of 1000 V/cm), which decreases analysis time and minimizes band diffusion. When separations are performed in the presence of electroosmotic flow (EOF), the plug-flow characteristics of EOF also contribute to high effciency. In contrast, the laminar flow properties of liquid chromatography increase resistance to mass transfer, reducing separation efficiency.

Because of its many advantages, CE shows great promise as an analytical tool in protein chemistry. In some cases it may replace HPLC and electrophoresis, but more often it is used in conjunction with existing techniques, providing a different separation selectivity, improved quantitation, or automated analysis. An anticipated benefit of performing protein separations in open-tubular capillaries was the reduced potential for surface interactions. In fact, this proved not to be the case; the high surface-to-volume ratio of the capillaries and the high surface activity of the fused silica capillary wall has proven to be a major problem in applying CE to protein separations. Much of the research in separation chemistries and capillary wall modifications has been directed toward improving CE performance in protein separations.

This book is designed to provide guidelines for separating proteins using CE. It has been compiled from a review of the published literature and from experience gained in the authors' labs over the last several years. It is written with an eye to the practical aspects of protein CE, and covers each of the major separation modes. Specific applications are included to illustrate the various methods as applied to different classes of proteins and different separation problems. A discussion of the basic principles of capillary electrophoresis is provided in Chapter 2 for readers less familiar with the field.

2

Principles and Practice
of Capillary Electrophoresis

This chapter provides a brief discussion of the separation mechanisms in capillary electrophoresis, a description of CE instrumentation, and some guidelines in selecting conditions for a CE separation. Readers interested in more detailed presentations of CE theory and practice may consult Refs. 1–8. Several general reviews of capillary electrophoresis have been published [9–11] as well as specific reviews of application of CE to protein analysis [12–15,16].

BASIC CONCEPTS

As the name implies, capillary electrophoresis separates species within the lumen of a small-bore capillary filled with an electrolyte. A schematic of a CE system is presented in Figure 1. The capillary is immersed in electrolyte-filled reservoirs containing electrodes connected to a high-voltage power supply. A sample is introduced at one end of the capillary (the inlet) and analytes are separated as they migrate through the capillary toward the outlet end. As separated components migrate through a section at the far end of the capillary, they are sensed by a detector and an electronic signal is sent to a recording device.

As in conventional gel electrophoresis, the basis of separations in capillary electrophoresis is differential migration of proteins in an applied electric field. The electrophoretic migration velocity v_{EP} will depend upon the magnitude of the electric field E and electrophoretic mobility μ_{EP} of the protein:

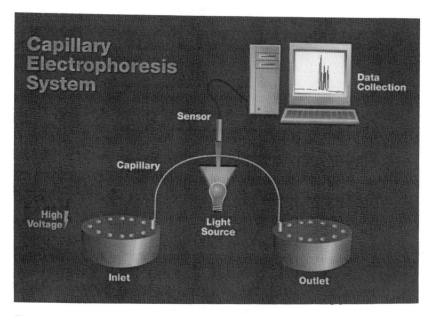

Figure 1 Schematic of a capillary electrophoresis system.

$$v_{EP} = \mu_{EP}E \tag{2-1}$$

In a medium of a given pH, mobility of a protein is given by the expression:

$$\mu_{EP} = \frac{q}{6\pi\eta r} \tag{2-2}$$

where q is the net charge of the protein, η is the viscosity of the medium, and r is the Stoke's radius of the protein. Since the Stoke's radius is related to molecular mass, mobility will increase inversely with molecular weight and directly with increasing charge. Mobility can be determined from the migration time and field strength by

$$\mu_{EP} = \left(\frac{l}{t}\right)\left(\frac{L}{V}\right) \tag{2-3}$$

where l is the distance from the inlet to the detection point (termed the effective length of the capillary), t is the time required for the analyte to reach the detection point (migration time), V is the applied voltage, and L is the total length of the capillary.

In contrast to most forms of gel electrophoresis, the velocity of an analyte in capillary electrophoresis will also depend upon the rate of electroendosmotic flow (EOF). This phenomenon is observed when an electric field is applied to a solution contained in a capillary with fixed charges on the capillary wall. Typically, charged sites are created by ionization of silanol groups on the inner surface of the fused silica. Silanols are weakly acidic, and ionize at pH values above about 3. Hydrated cations in solution associate with ionized SiO^- groups to form an electrical double layer, a static inner Stern layer close to the surface and a mobile outer layer, termed the Helmholtz plane. Upon application of the field, hydrated cations in the outer layer move toward the cathode, creating a net flow of the bulk liquid in the capillary in the same direction (Figure 2). The rate of movement is dependent upon the field strength and the charge density of the capillary wall. The population of charged silanols is a function of the pH of the medium, so the magnitude of EOF increases directly with pH until all available silanols are fully ionized (Figure 3). The velocity v_{EOF} of electroendosmotic flow can be expressed as

$$v_{EOF} = \mu_{EOF}E \tag{2-4}$$

where μ_{EOF} is the electroendosmotic mobility, defined as:

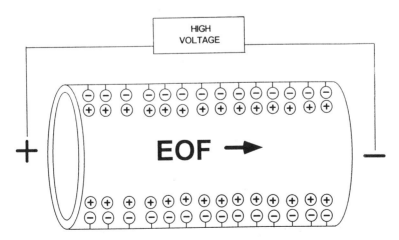

Figure 2   Electroendosmotic flow in an uncoated fused silica capillary.

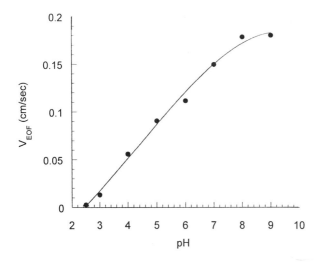

Figure 3   Magnitude of electroendosmostic flow as a function of pH.

$$\mu_{EOF} = \frac{\varepsilon\xi_W}{\eta} \qquad (2\text{-}5)$$

where, ξ_W is the zeta potential of the capillary wall, ε is the dielectric constant, and η is the viscosity of the medium. Electroendosmotic mobility can be determined experimentally by injecting a neutral species and measuring the time t_{EOF} when it appears at the detection point:

$$\mu_{EP} = \left(\frac{l}{t_{EOF}}\right)\left(\frac{L}{V}\right) \qquad (2\text{-}6)$$

The apparent velocity v_{APP} of a protein in an electric field will therefore be the combination of its electrophoretic velocity and its movement in response to EOF:

$$v_{APP} = v_{EP} + v_{EOF} = (\mu_{EP} + \mu_{EOF})E \qquad (2\text{-}7)$$

In the presence of electroendosmotic flow, proteins which possess net positive charge will migrate faster than the rate of EOF, proteins which are isoelectric will be carried toward the cathode at the rate of EOF, and anionic proteins will migrate toward the cathode at a rate which is the difference between their electrop[...] v_{EOF}. If the magnitude of EOF is sufficiently g[...] gardless of their charge state will migrate past the detection point. In this regard, performing separations in the presence of EOF is highly desirable. However, acheiving reproducible separations requires that EOF is constant, and this in turn requires that the surface character- istics of the capillary wall remain constant from run to run. Proteins are notorious for interacting with silica surfaces and changing the level of EOF. A great variety of capillary surface treatments and buffer additives have been developed for reducing protein adsorption and controlling EOF, and these are described in detail in Chapter 3.

Separation efficiency N in capillary electrophoresis is given by the following expression:

$$N = \frac{\mu_{APP} V}{2D} \tag{2-8}$$

where D is the diffusion coefficient of the analyte in the separation medium. This predicts that efficiency is only diffusion-limited and increases directly with field strength. Capillary electrophoresis sepa- rations are usually characterized by very high effiency, often as high as several hundred thousand plates. Efficiency in CE is much higher than in HPLC because there is no requirement for mass transfer be- tween phases and because the flow profile in EOF-driven systems is flat (approximating plug flow) in contrast to the laminar-flow pro- files characteristic of pressure-driven flow in chromatography col- umns (Figure 4). According to the above expression, proteins should exhibit excellent separation efficiencies since they possess low dif- fusion coefficients relative to small molecules. Resolution R in CE is defined as

$$R = \frac{1}{4} \left(\frac{\Delta\mu_{EP} N^{1/2}}{\mu_{EP} + \mu_{EOF}} \right) \tag{2-9}$$

This implies that resolution will be greatest when μ_{EP} and μ_{EOF} are of similar magnitude but of opposite sign; however, high resolution will be at the expense of analysis time. It is also evident from Eq. (2-9) that, although Eq. (2-8) suggests use of higher field strengths is the

A

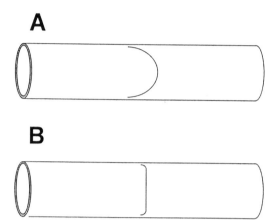

B

Figure 4 Velocity profiles in laminar flow (A) and electroendosmotic flow (B).

most direct route to high efficiency, a doubling of voltage yields only a 1.4-fold increase in resolution, at the expense of Joule heat.

BAND BROADENING IN CAPILLARY ELECTROPHORESIS

Band broadening and resultant reduction in resolution can arise from several contributing factors. If the total band broadening is expressed as plate height H, the contributions to band broadening due to initial zone width, diffusion and electrodispersion, Joule heating, and adsorption can be expressed qualitatively [17] as:

$$H = h_{inj} + h_{diff+cond} + h_{joule} + h_{abs} \qquad (2\text{-}10)$$

Initial Zone Width (h_{inj})

Best resolution will always be obtained by keeping initial sample zone as small as possible. The starting zone length should not exceed 5% of the total capillary length. In electrokinetic injection, injection zone length L_{inj} can be estimated from the injection time T_{inj}, the injection and separation field strengths E_{inj} and E_{sep}, and the migration time T_m of the peak of interest:

$$L_{inj} = \frac{(L_{eff})(E_{inj})(T_{inj})}{(E_{sep})(T_m)} \qquad (2\text{-}11)$$

Sample zone sharpening can be achieved by preparing the sample in an electrolyte of lower conductivity than the analysis buffer [18–20]. Under these conditions there is a discontinuity in field strength at the sample/buffer boundary such that ions migrating rapidly from a region of higher field strength become focused at the boundary. This focusing or stacking effect not only produces narrow zones for increased resolution but increases zone concentration for enhanced sensitivity. Sensitivity enhancement is directly proportional to the ratio of sample to buffer conductivity.

In displacement injection, the injection zone volume V_{inj} in cubic centimeters can be estimated from the Poiseuille equation:

$$V_{inj} = \frac{\pi p t r^4}{8L\eta} \qquad (2\text{-}12)$$

where p is the pressure in dynes/cm^2, t is the injection time in seconds, r is the capillary radius in centimeters, L is the total length of the capillary in centimeters, and η is the viscosity of the electrolyte in poise. Pressure in psi can be converted to dynes/cm^2 using the value of 68947.6 dynes/cm^2/psi. Stacking effects will also be observed in displacement injection if the sample electrolyte and analysis buffer differ in conductivity, but the effect is less pronounced than seen in electrokinetic injection. An alternative method of stacking employs differences in sample and buffer pH (12). The sample is prepared in an alkaline solution in which analyte polypeptides are negatively charged; the sample is pressure-injected into a capillary containing buffer at a lower pH. When voltage is applied, sample ions stack at the anodic end of the sample zone. Following the stacking phase, the sample electrolyte diffuses away from the zone, and polypeptides become positively charged and migrate toward the cathode.

Diffusion and Electrodispersion ($h_{diff+cond}$)

Band broadening due to axial diffusion should be reduced by shortening analysis time, i.e., by operating at high field strengths with short capillaries. On the other hand, higher field strengths will

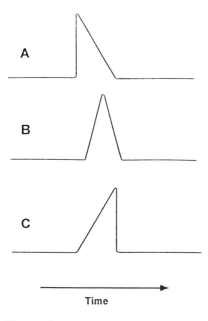

Figure 5 Peak asymmetry due to electrodispersion. (A), conductivity of zone greater than conductivity of electrolyte; (B) conductivity of zone equal to conductivity of electrolyte; (C) conductivity of zone less than conductivity of electrolyte.

generate more Joule heat, which increases diffusion rates. Diffusive band broadening is minimal for polypeptides, which possess small diffusion coefficients. Electrodispersive band broadening arises from conductivity differences between the zone and the background electrolyte. If this difference is large, diffusion at one boundary of the zone is neglible due to the zone sharpening effect caused by the discontinuity in field strength. In this case, the peak will be asymetric: the peak will exhibit fronting or tailing depending on whether conductivity of the zone is greater or less than that of the background electrolyte (Figure 5). This phenomenon is readily observable when analyzing highly charged small molecules such as inorganic ions, and satisfactory peak shapes can only be obtained by carefully matching sample and electrolyte conductivities. The effect is much less noticeable in most protein separations, except at high sample concentrations.

Joule Heating (h_{joule})

When an electric field is applied to a capillary containing an electrolyte, Joule heat is generated uniformly across the circumference of the tube. Since heat can be removed only at the margin of the tube, a temperature gradient exists across the radius of the tube (Figure 6). As noted in Eq. (2-2), mobility is inversely related to viscosity, which decreases with temperature. Mobility increases approximately 2.5% for each degree rise in temperature. Therefore, the temperature gradient creates a mobility gradient across the tube radius which contributes to band broadening. It should also be noted that, in addition to this mobility gradient, Joule heat can compromise resolution by increasing diffusion and convection; these are minimized by the use of viscous buffers.

The small internal diameters of fused silica capillaries have been the key to high-resolution CE separations. The smaller the capillary bore, the greater the surface-to-volume ratio and the more efficiently heat is removed from the tube. The optimal capillary inside diameter for most applications is 50–75 µm. Capillaries with smaller diameters are subject to plugging, and the high surface-to-volume ratio increases the risk of protein adsorption. Capillaries with larger internal diameters may exhibit significant loss in resolution due to thermal effects, and may require operation at low field strengths or with low-conductivity buffers.

Adsorption (h_{abs})

Protein-wall interactions have been the greatest obstacle to achieving satisfactory resolution and reproducibility in capillary electro-

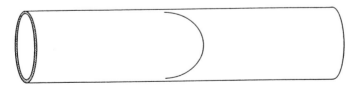

Figure 6 Profile of migration velocity in a capillary due to Joule heat. Migration velocity increases approximately 2.5% for each degree rise in temperature; Joule heat is removed only at the capillary wall, resulting in a heat (and velocity) gradient across the capillary radius.

phoresis, particularly in zone electrophoresis and isoelectric focusing. Approaches to minimizing or eliminating protein adsorption include use of buffer additives and dynamic or covalently coupled capillary coatings. These various approaches are discussed extensively in Chapter 3.

CAPILLARY ELECTROPHORESIS INSTRUMENTATION

A wide variety of commercial CE instruments is available, from simple modular systems consisting of a power supply, detector, and injection device to fully integrated automated systems under computer control. This section focuses on the features of automated CE instrumentation.

Power Supply

Power supplies capable of delivering constant voltage at high precision up to 30 kV are standard throughout the industry. Most systems offer, in addition, constant current operation at up to 300 µA; constant current operation may be desirable in systems without adequate temperature control or for special applications. High voltage is applied at the capillary inlet, with the outlet (detector) at ground potential. When separations are performed in uncoated fused silica capillaries in the presence of EOF, the inlet (high-voltage) electrode is the anode and EOF carries analytes toward the cathode ("normal" polarity). Separations performed in coated capillaries without EOF may require "reversed" polarity. In some separations the direction of EOF may be reversed by using capillaries coated with positively charged polymers or by using osmotic flow modifying additives in the background electrolyte. In these cases reversed polarity is also used. All commerical instruments have reversible polarity, although for frequent switching among different applications, polarity reversal through software is more convenient than manual reversal.

Injection

Sample injection in CE requires the introduction of very small amounts of analyte at the capillary inlet with high precision. All commercial instruments offer electromigration injection and at least one type of displacement injection.

Electromigration is the simplest injection method in CE; the capillary inlet is immersed in the sample solution and high voltage is applied for a brief period (typically a few seconds). If no electroosmotic flow is present, sample ions enter the capillary by electophoretic mobility alone. If EOF is present, sample ions will be introduced by a combination of electrophoretic mobility and electroosmotic flow; this mode is generally termed electrokinetic injection.

Electrophoretic injection offers two advantages. First, if electrophoretic injection is performed in the absence of EOF, only species of like charge will enter the capillary. This enables discrimination against compounds of opposite charge, simplifying the separation problem. Second, zone sharpening can be achieved using the stacking principle described above. Unfortunately, these advantages are countered by two major limitations. Since sample ions enter the capillary based on mobility, low-mobility ions will be loaded to a lesser extent than high-mobility ions. More importantly, presence of nonanalyte ions in the sample will reduce injection efficiency, so electrophoretic injection is very sensitive to the presence of salts or buffers in the sample matrix. The disadvantages of electrophoretic injection argue against its use in routine analysis except in cases where displacement injection is not possible, e.g., in capillary gel electrophoresis (CGE). Electrokinetic injection suffers from the further disadvantage that many sample matrices contain components such as proteins which adsorb to the capillary wall and change the magnitude of EOF.

Displacement injection is usually the preferred method since analyte ions are present in the sample zone in proportion to their concentration in the bulk sample, and injection efficiency is less sensitive to variations in sample ionic strength. However, it should be noted that the presence of high salt can affect detector response with displacement injection, and variations in the sample viscosity (due to temperature variations or the presence of viscosity-modifying components) can affect displacement injection efficiency.

Two modes of displacement injection have been employed in commercial CE instruments: application of positive pressure at the capillary inlet and application of vacuum at the capillary outlet. The former method can employ pressurization of the sample headspace by gas (pressure injection) or application of hydrostatic pressure by elevating the capillary inlet relative to the capillary outlet (gravity injection). Gravity injection has been reported to provide reproducible injection of very small sample zones. On the other hand, pressure injection can allow the flexibility of introducing larger sample zones; this can be an advantage when using injection pneumatics to introduce a chiral selector or to perform on-column concentration. Moderate injection pressures can be an advantage when injecting samples into capillaries containing analysis buffers with viscous agents such as sieving polymers. High precision in displacement injection is achieved by integrating the pressure signal during the injection process and injecting to a constant time-pressure product; this compensates for variations in seal compliance and sample-to-sample headspace.

Capillary Temperature Control

Temperature control of the capillary environment is essential for attaining satisfactory reproducibility. Inadequate temperature control results in variable migration times. In CE, peak area depends upon the residence time of the component in the detector light path and therefore is dependent upon migration velocity. If migration times vary because of inadequate temperature control, peak area precision will be poor. Control of capillary temperature above or below ambient temperature may be desirable in special applications, for example, in performing kinetic studies, on-column enzyme assays, or in the study of protein folding. Operation at higher temperatures in capillary zone electrophoresis will decrease analysis time and may improve peak shape, although at high temperatures the risk of protein denaturation and precipitation is increased. Operation at lower temperatures generally has no advantage in most modes of capillary electrophoresis.

Capillary temperature control can be achieved by forced air or nitrogen convection or by a circulating liquid coolant. Forced air control permits the use of free-hanging capillaries but is less efficient. Liquid cooling requires that the capillary be enclosed in a sealed cartridge, but the cartridge format provides for automatic alignment of the capillary in the detector light path and reduces time required for capillary installation when changing methods.

The effectiveness of capillary thermostatting can be determined by variation in current as a function of voltage. According to Ohm's law, this should be a linear relationship, and deviation from linearity in an Ohm's law plot is indicative of poor efficiency in heat dissipation by the capillary temperature control system.

Detectors

Absorbance

As in HPLC, absorbance detection is used in the vast majority of CE applications, and all commercial CE systems employ UV or UV-Vis absorbance as the primary mode of detection. The simplest approach is the use of line-source lamps or continuum-source lamps with wavelength selection by filters. Better flexibility is obtained using a continuum (e.g., deuterium lamp) source with wavelength selection by a grating monochromator. The low output of a deuterium lamp above 360 nm limits sensitivity in the visible range and a secondary tungsten source provides best performance for visible-wavelength detection.

All commercial CE absorbance detectors employ on-tube detection: a section of the capillary itself is used as the detection cell. This permits detection of separated zones with no loss in resolution. Most capillaries used for CE are coated ~~with~~ with a polymer (usually polyimide) which protects the fused silica capillary and provides it with mechanical stability. Since the polymer is not optically transparent, it must be removed from the detection segment to form a "window," and this window must be accurately positioned in the optical path to achieve good sensitivity. This segment of bare capillary is very fragile, and is subject to breakage during manipulation and installation of the capillary. Capillaries with a UV-transparent coating, which eliminate this problem, are available from Polymicro Technologies; however, the coating is not resistant to some coolants (e.g., fluorinated hydrocarbons) used in liquid-cooled CE systems.

In on-tube detection, the internal diameter of the capillary forms the detection light path. In accordance with Beer's law, the sensitivity of a concentration-sensitive detector is a direct function of the length of the light path. Therefore, in comparison to an HPLC detector with a 1 cm path length, detector signal strength should be reduced 200-fold in a CE system equipped with a 50 μm i.d. capillary. Concentration sensitivity can be improved by employing focusing lenses to collect light at the capillary lumen, by detecting at low wavelengths (where most analytes have greater absorbance), and by using sample focusing techniques during the injection process. However, even under ideal conditions, the concentration limit of detection (CLOD) is about $10^{-6}M$.

Several commercial CE systems incorporate scanning absorbance detectors. Scanning detection enables on-the-fly acquisition of spectra as analytes migrate through the detection point; this information can assist in the identification of peaks based on spectral patterns, in detection of peak impurities by variation in spectral profiles across a peak, or in determation of the absorbance maximum of an unknown compound. Two different designs are used to accomplish scanning detection in CE instruments. In photodiode array (PDA) detectors, the capillary is illuminated with full-spectrum source light; the light passing through the capillary is dispersed by a grating onto an array of photodiodes each of which samples a narrow spectral range. In fast-scanning detectors, monochromatic light is collected from the source using a moveable grating and slit assembly and directed to the capillary; light transmitted by the capillary is detected by a single photodiode. Scanning is accomplished by rapidly rotating the grating through an angle to "slew" across the desired spectral range.

Fluorescence

Fluorescence detection offers the possibility of high sensitivity, and in the case of complex samples, improved selectivity. However, this mode of detection requires that the analyte exhibit native fluorescence or contain a group to which a fluorophore can be attached by chemical derivatization. The number of compounds that fall into the former category are small, and while many analytes contain derivatizable groups (e.g., amino, carboxyl, hydroxyl), most derivatization chemistries are limited by one or more disadvantages (slow reaction kinetics, complicated reaction or cleanup conditions, poor yields, interference by matrix components, derivative instability, interference by reaction side products, or unreacted derivatizing agent). Since only two of the protein amino acids exhibit native fluorescence, fluorescence detection of proteins usually requires derivatization. However, most proteins possess multiple reactive sites, and incomplete derivatization yields a family of products varying in the number of fluorophores. The reactive sites are usually side-chain amino groups, and the derivatization products (which vary in mass and charge) may be resolved into multiple peaks or migrate as a single broad peak.

When compared to fluorescence detectors for HPLC, the design of a fluorescence detector for CE presents some technical problems. In order to obtain acceptable sensitivity, it is necessary to focus sufficient excitation light on the capillary lumen. This is difficult to achieve with a conventional light source, but is easily accomplished using a laser. The most popular source for laser-induced fluorescence (LIF) detection is the argon ion laser, which is stable and relatively inexpensive. The 488 nm argon ion laser line is close to the desired excitation wavelength for several common fluorophores. The CLOD for a laser-based fluorescence detector can be as low as $10^{-12}M$.

On-line Coupling with Mass Spectrometry (MS)

With the increasing need to obtain absolute identification of separated components and the gradual price reduction of mass spectrometers, there is a growing demand for direct coupling of CE with MS instruments. The most frequent configuration is introduction of the capillary outlet into an electrospray interface (ESI) coupled to the mass spectrometer. In this configuration, the outlet electrode of the CE is eliminated and the MS becomes the ground. Since the volumetric flow out of the capillary is neglible or nil, separated components are usually transported from the capillary to the electrospray using a liquid sheath flow. The major limitation in CE-ESI/MS is the requirement for volatile buffers. This narrows the choice of CE separation modes and resolving power.

PREPARATIVE CAPILLARY ELECTROPHORESIS

Because of its high resolving power, CE is often considered for micropreparative isolation of proteins. Many of the commercially available CE systems have the capability for automatic fraction collection. However, the desirability of using CE as a preparative tool has to be carefully weighed against the problems encountered in fraction collection. When using narrow-bore (e.g., 50 μm i.d.) capillaries, the volume injected into the capillary is quite small (typically a few nanoliters). Unless the analyte is in very high concentration, recovery of sufficient material will require repetitive injections of the same sample. In this case, the run-to-run migration times must be highly reproducible to ensure accurate collection of the analyte peak. Also, the recovered analyte must be stable under the collection conditions for the time required to collect the desired amount of material (often several hours). An alternative strategy is the use of larger-diameter capillaries (≥ 75 μm). However, thermal effects may compromise resolution, and low voltages or low-conductivity buffers may be necessary to prevent excessive heating.

CAPILLARY ELECTROPHORESIS SEPARATION MODES

One of the major advantages of CE as a separation technique is the wide variety of separation modes available. Analytes can be separated on the basis of charge, molecular size or shape, isoelectric point, or hydrophobicity. The same CE instrument can be used for zone electrophoresis, isoelectric focusing, sieving separations, isotachophoresis, and chromatographic techniques such as micellar electrokinetic chromatography and capillary electrokinetic chromatography. This section provides a brief description of each separation mode; zone electrophoresis, isoelectric focusing, and sieving are the primary modes used for protein separations and these are treated in detail below.

Capillary Zone Electrophoresis (CZE)

In CZE, the capillary, inlet reservoir, and outlet reservoir are filled with the same electrolyte solution. This solution is variously termed background electrolyte, analysis buffer, run buffer, or mobile phase. The last term is commonly used by practioners experienced in chromatography, but it is not the best choice for CE since there will be no bulk liquid movement if there is no EOF. In CZE the sample is in-jected at the inlet end of the capillary and components migrate to-ward the detection point according to their mass-to-charge ratio by the principles outlined above. It is the simplest form of CE and the most widely used, particularly for protein separations. Capillary zone electrophoresis is covered in Chapter 4.

Capillary Isoelectric Focusing (CIEF)

Capillary IEF is similar in concept to conventional gel IEF; a stable pH gradient is formed in the capillary using carrier ampholytes, and proteins become focused in the gradient at their isoelectric points. The major difference in performing IEF in the capillary format is the requirement for mobilizing focused protein zones past the detection point. Capillary IEF is described in Chapter 5.

Capillary Sieving Techniques

Sieving techniques are required for separation of species which have no difference in mass-to-charge ratio. This includes native proteins composed of varying numbers of identical subunits, protein aggre-gates, and SDS-protein complexes. Sieving systems include cross-linked or linear polymeric gels cast in the capillary, or replaceable polymer solutions. Sieving techniques are described in Chapter 6.

Isotachophoresis (ITP)

As a separation techique, ITP resolves proteins as contiguous zones which migrate in order of mobility. The sample is injected into the capillary between a leading buffer (with ion mobility greater than that of all protein components) and a terminating buffer (with ion mobility less than that of all protein components). Zones migrate at equal velocity toward the detection point where they are detected as steps with zone length proportional to concentration. When UV-trans-parent spacers are added to the sample, ITP zones may appear as isolated peaks and the detector trace will resemble a CZE electro-pherogram. Isotachophoresis is rarely used as a separation method for proteins, but is occasionally used as a preconcentration technique, as described in Chapter 4.

Micellar Electrokinetic Chromatography (MEKC)

As the name implies, MEKC is a chromatographic technique in which samples are separated by differential partitioning between two phases. The technique is usually performed in uncoated capillaries under alkaline conditions to generate a high electroendosmotic flow. The background electrolyte contains a surfactant at a concentration above its CMC (critical micelle concentration); surfactant monomers are in equilibrium with micelles. The most widely used MEKC system employs sodium dodecylsulfate (SDS) as the surfactant. The sulfate groups of SDS are anionic, so both surfactant monomers and micelles have electrophoretic mobility counter to the direction of EOF. Sample molecules will be distributed between the bulk aqueous phase and the micellar phase depending upon their hydrophobic character (see Figures 7, 8). Hydrophilic neutral species with no affinity for the micelle will remain in the aqueous phase and reach the detector in the time required for EOF to travel the effective length of the column. Hydrophobic neutral species will spend varying amounts of time in the micellar phase depending on their hydrophobicity, and their migration will therefore be retarded by the anodically moving micelles. Charged species will display more complex interactions since they have the potential for electrophoretic migration and electrostatic interaction with the micelles in addition to hydrophobic partitioning. The selectivity of MEKC can be expanded with the introduction of chiral selectors for chiral surfactants to the system. MEKC is used almost exclusively for small molecules such as drugs and

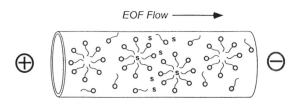

Figure 7 Schematic of micellar electrokinetic chromatography. Sample molecules (S) are distributed between the bulk mobile phase and surfactant micelles. Note that surfactant molecules are in equilibrium between free monomers and micelles.

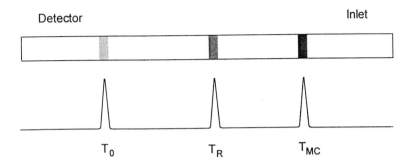

Figure 8 Profile of an MEKC chromatogram (below) and its representation in the capillary (above). The unretained peak (T_0) represents the time required for EOF-pumped mobile phase to traverse the distance from the capillary inlet to the detection point. The micellar marker peak (T_{MC}) represents the time required for micelles to traverse the same distance, and T_R represents the time for an analyte which is partially included in the micellar phase to reach the detector.

metabolites; it has been used occasionally for peptides. The MEKC mechanism is not applicable for proteins since they are too large to partition into a surfactant micelle, and bind surfactant monomers tenaciously to form SDS-protein complexes.

Capillary Electrokinetic Chromatography (CEC)

Like MEKC, CEC is a chromatographic technique performed with CE instrumentation. It employs fused silica capillaries packed with 1.5–5 μm microparticulate porous silica beads, usually derivatized with a hydrophobic ligand such as C18. Mobile phases are similar to those used for conventional reversed phase HPLC, e.g., mixtures of aqueous buffers and an organic modifier such as acetonitrile. The silica surface of the derivatized beads has sufficient densities of ionized silanol groups to generate a high electroendosmotic flow when a voltage is applied to the system, pumping mobile phase through the column. Since EOF is pluglike rather than laminar in nature, efficiencies in CEC can be much higher than in HPLC. Like MEKC, CEC is used primarily for small molecules.

3
Detection of Proteins

As capillary electrophoresis is increasingly applied to analysis of proteins present at trace levels in biological materials or to determination of impurities and degradation products in formulations of protein therapeutics, detection sensitivity becomes a limiting factor in obtaining quantitative information. Several strategies have been pursued to increase detection sensitivity in capillary electrophoresis of proteins.

ABSORBANCE DETECTION

In liquid chromatography, absorbance detection at 280 nm is typically used to monitor the separation. The detector signal at this wavelength is due to the absorbance of aromatic residues in the polypeptide, e.g., tryptophan, tyrosine, and (to a lesser extent) phenylalanine. This detection wavelength is sufficiently selective for polypeptides so that nonprotein interferences in the sample are not detected, while the pathlength of HPLC absorbance detectors (nominally 1 cm) provides sufficient signal for satisfactory sensitivity. However, reduction of the detector pathlength to 25–75 μm in on-tube detection cuts detector signal by a factor of 100–400 in capillary electrophoresis. Therefore, detection at 280 nm rarely provides sufficient signal for satisfactory sensitivity. Instead, detection at 200 nm is typically employed where proteins exhibit 50- to 100-fold greater absorbtivity (Table 1). In spite of this sensitivity gain, detection in the low UV region may be inadequate for many applications in capillary electrophoresis of proteins.

Table 1 Protein Response in Milliabsorbance Units (mAU) at UV
Wavelengths[1]

Protein	200 nm	220 nm	250 nm	280 nm
Bovine serulm albumin	124	34	1.0	1.8
Ovalbumin	83	20	0.5	1.1
Myoglobin	104	32	0.9	1.9
Lysozyme	131	58	4.0	8.3
Cytochrome C	111	33	3.8	4.6

[1] Protein solutions (1 mg/ml in 5 mM sodium phosphate, pH 2.5) injected by pressure (5 psi*s) into a 24 cm × 50 μm i.d. polyAAEE-coated capillary and separated at 10 kV at 20°C using 50 mM sodium phosphate (pH 2.5). Peak height in mAU determined using multiwavelength detection on a BioFocus 3000TC CE system.

Precolumn derivatization techniques for proteins have been developed to increase detection sensitivity in CE. In this approach, functional groups on the protein are reacted with a UV-absorbing species to form a detectable product. Precolumn derivatization is widely used for analysis of amino acids, in which one of a variety of derivatizing agents such as o-phthalaldehyde (OPA), phenylisothiocyanate (PITC), or naphthalene dicarboxaldehyde (NDA) is reacted with the α- or side-chain amino groups and the derivatives are separated by reversed phase HPLC. Precolumn derivitization can also be applied to proteins, but there are inherent limitations to derivatizing proteins prior to CE separation. Amino acids have one or two reactive sites and the derivatization process typically yields a single product. In contrast, proteins have multiple reactive sites depending upon the number of residues with reactive groups and their accessibility under the derivatization conditions. In CZE, mobility is a function of the mass/charge ratio of the protein. Since the side-chain amino groups are major contributors to the total charge of the protein, variation in the population of derivatized side chains produces a multplicity of derivatization products with varying mobilities. This results in peak broadening or, in extreme cases, resolution of discreet peaks for different derivatives of the same protein. It is possible to reduce generation of multiple dertivatives by using mild derivatization conditions so that only the most reactive sites are tagged, or by using very aggressive derivatization conditions which exhaustively label all possible derivatizable groups. In addition to these specific considerations in precolumn derivitization of proteins for capillary electrophoresis, there are the more general concerns of interference from the derivatization reagent or its hydrolysis products, interference from other sample components (e.g., amino acids, amine-containing buffer constituents), instability of derivatives, and matrix affects on derivative yield. Guzman et al. [21] demonstrated precolumn derivitization of leukocyte A interferon and a monoclonal antibody for enhanced UV detection. The proteins were reacted with fluorescamine and separated under CZE conditions. Derivitization improved UV response at 280 nm by 20-fold, but significantly increased protein migration time and peak width.

Use of precolumn derivatization prior to sieving separations of SDS-protein complexes circumvents some of the problems encountered in CZE separations of derivatized proteins. In this application, separations are based on size rather than mass/charge ratio. After complexation of proteins with SDS, the anionic surfactant dominates the charge of the protein, swamping any contribution from the amino acid side chains. The complexed proteins then have approximately constant mass/charge ratios independent of size, and separation of the random-coil complexes is achieved by the sieving matrix. In addition, excess labeling reagent migrates ahead of proteins since it is below the sieving threshold of the medium. Gump and Monnig [22] investigated three precolumn derivatizing agents (OPA, NDA, and fluorescamine) for enhancing detection sensitivity of SDS-protein complexes separated by an entangled polymer sieving system. Comparing absorbance detection at several wavelengths for labeled vs. unlabeled proteins, signal enhancements of up to 20-fold were observed at 280 nm while enhancement factors were more modest at lower wavelengths. The UV response of derivatized proteins was protein-dependent, reflecting varying numbers of derivatizable groups. Separation efficiency was reduced substantially for some proteins, reflecting zone broadening caused by multiplicity of labeled states. The observed protein molecular-weight values were apparently not affected by the derivatization process since they agreed well with published values.

FLUORESCENCE DETECTION

The high sensitivity and selectivity of fluorescence detection make this the obvious choice for improving detection of proteins. Four approaches have been used: direct detection of intrinsic protein fluoresence, indirect fluorescence detection, precolumn derivatization, and postcolumn reaction detection.

Direct Detection of Intrinsic Protein Fluorescence

Direct detection of proteins by the intrinsic fluorescence of tryptophan and tyrosine residues provides enhanced sensitivity compared to absorbance detection without the complexity of pre- or postcolumn derivatization. However, the optimal excitation wavelengths for these amino acids is in the 270–280 nm range which is beyond the capability of the low-cost lasers used in commercial LIF detectors. At this time, detection of intrinsic protein fluorescence is limited to higher-cost detectors assembled in research labs. Swaile and Sepaniak [23], using a home-built instrument, detected instrinsic fluorescence of proteins by excitation at 257 nm with an argon ion laser operated at 514 nm with frequency-doubling using a harmonic generator. The concentration limit of detection for conalbumin was $2.5 \times 10^{-8}\,M$ for this system. Lee and Yeung [24] were able to achieve a CLOD of 5×10^{-10} for the same protein using the 275.4 nm line of an argon ion laser producing deep UV light, at the expense of higher cost of the laser. Detection of intrinsic protein fluorescence is limited by the variations in fluorescence yield due to variable numbers of aromatic residues and the influence of amino acid microenvironment on fluorescence quantum efficiency.

Indirect Fluorescence Detection of Proteins

A simple alternative to direct detection of instrinsic protein fluorescence detection is the technique of indirect fluorescence detection proposed by Kuhr and Yeung [25]. In this approach, the electrophoresis buffer contains a fluorescent anion which produces a high background fluorescence signal. Nonfluorescent analyte anions displace the fluorescent species, producing a zone of reduced signal. This technique, similar in concept to the indirect absorbance mode used for inorganic ions and organic acids [26] enables universal detection of analyte ions which possess neither effective chromophores or fluorophores.

Sensitivity in indirect fluorescence detection is determined by the dynamic reserve (ratio of signal intensity to signal fluctuation, S/N) and the displacment ratio. Sensitivity is increased by reducing the concentration of the fluorophore, but operation at low buffer ionic strengths increases the potential for wall interactions and varying EOF. Using a coated capillary and salicylate as the background fluorophore, a detection limit of 100 amol was demonstrated for lysozyme (excitation of salicylate at 325 with a stabilized HeCd laser, collection of emission at 405.1 nm).

Precolumn Derivatization for Fluorescence Detection

The precolumn derivatization procedure described above for absorbance detection can also be applied to fluorescence detection and, in fact, many of the reagents used (OPA, NDA, fluorescamine) are also highly fluorescent. Most of the caveats cited for precolumn derivatization also apply to fluorescence detection, e.g., interference from derivatization reagents and side products, matrix-sensitive derivitization chemistries, altered migration behavior of derivatives, and broad or multiple peaks arising from multiple derivitization products. Loss of efficiency and resolution of multiple species has been observed in CZE separation of proteins following precolumn derivatization with fluorescein isothiocyanate [23] and OPA [27].

Swaile and Sepaniak [23] developed an on-column protein-labeling method in which fluorescent hydrophobic probes were added to the electrophoresis buffer. Intercalation of the probes into the hydrophobic domains of the proteins increased their fluorescence quantum efficiency with resulting enhancement of fluorescence signal intensity. Using 1-anilinonaphthalene-8-sulfonate or 2-p-toluidinonaphthalene-6-sulfonate as probes and LIF detection with a HeCd laser (excitation at 325 nm, emission collected at 500 and 450 nm, respectively), detection limits intermediate between absorbance and intrinsic fluorescence detection were obtained for conalbumin.

Separation of proteins as SDS complexes in sieving media avoids some of the problems of precolumn fluorescent labeling, as noted above for absorbance detection. Gump and Monnig [22] evaluated LIF detection of SDS-protein complexes prelabeled with NDA, fluorescamine, or OPA using the 457.9, 363.8, or 351.1 nm lines of an argon ion laser, respectively. To achieve satisfactory derivative yields, the derivitization reaction had to be performed prior to SDS complexation for fluorescamine and NDA, while this was not possible with OPA due to derivative instability. Detection limits were comparable for OPA and NDA (e.g., 3–300 amol injected) but somewhat higher for fluorescamine. Wise et al. [28] used 4-fluoro-7-nitrobenzofurazan and 4-chloro-7-nitrobenzofurazan as precolumn derivatization reagents prior to separation of SDS-protein complexes in an entangled polymer sieving system. Fluorescence detection used an argon ion laser with excitation at 488 nm and collection of emission at 520 nm. Optimization of the reaction conditions yielded a rapid labeling procedure which provided separation efficiencies comparable to unlabeled proteins and detection limits 20-fold lower than UV detection. Molecular mass estimates of labeled proteins were within 5% of published values.

Recently Pinto et al. [29] developed a precolumn procedure for labeling proteins with a fluorogenic reagent, 5-furoylquinoline-3-carboxaldehyde (FQ), which provided excellent sensitivity and minimized the band-broadening caused by multiple derivatization products produced by reaction of FQ with side-chain lysines (Figures 1 and 2). Following derivatization, separations were carried out in the presence of 5 mM sodium dodecylsulfate; the submicellar concentration of the surfactant apparently masked the mobility differences of the multiple derivatization products, producing sharp peaks for FQ-protein derivatives with efficiencies of over 190,000 theoretical plates (Figures 1 and 2). The authors demonstrated assay detection limits (i.e., concentration detection limits for unlabeled protein taken through the derivatization and separation process) of 1×10^{-11} M for conalbumin. Limitations of this technique include variable protein sensitivity due to differing numbers of lysine residues and the masking of the mobility differences of protein glycoforms. This study was performed using a sheath-flow argon ion LIF fluorescence detection system; sensitivity using a commercial on-column argon ion LIF system was about 100× poorer.

Postcolumn Reaction Detection

Formation of derivatives after the separation process avoids the problems described above for precolumn derivatization. A typical postcolumn reactor for CE involves coupling the separation capillary

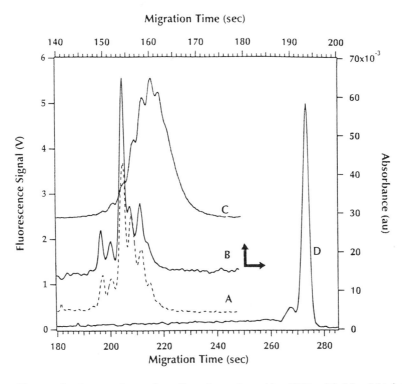

Figure 1 Comparison of ovalbumin analyzed by CZE with 25 mM tricine, pH 8.0 (A–C) or 25 mM tricine + 5 mM SDS (D) with various detection modes. (A) albumin labeled with FQ for 15 s, LIF detection; (B) unlabeled ovalbumin, UV absorbance detection; (C) FQ-labeled ovalbumin, 10 min reaction, LIF detection; (D) FQ-labeled ovalbumin, 1 min reaction. Electropherogram B is plotted vs. top x-axis and right y-axis. (Reproduced from Ref. 29 with permission.)

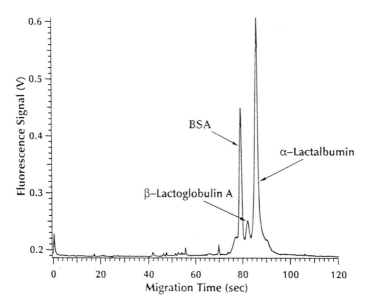

Figure 2 Separation of FQ-labeled bovine serum albumin, β-lactoglobu-lin, and α-lactalbumin labeled at a concentration of 5, 10, and 15 nM, re-spectively (60 s reaction time, 5 s × 2 kV injection, 30 cm × 50 μm capil-lary, separation at 700 V/cm in 2.5 mM borax + 5 mM SDS). (Reproduced from Ref. 29 with permission.)

with a reagent capillary and a reactor capillary through a mixing tee. The derivatization reagent is pumped into the tee, mixes with analytes emerging from the separation capillary and reacts with proteins to form fluorescent products during transit through the reactor capil-lary to the detector. The postcolumn reactor must be designed so that the derivatization reaction is complete by the time analytes reach detection point and there is minimal band broadening during reagent mixing and transport to the detector. This requires careful optimiza-tion of reagent flow (pumping rate, capillary diameter, and capillary length) and geometry of the mixing tee.

Nickerson and Jorgenson [27] constructed a mixing tee in which the etched outlet end of a separation capillary was inserted coaxially into the inlet end of the reactor capillary. o-Phthalaldehyde was de-livered by helium gas pressure, and OPA-protein derivatives were detected by fluorescence using a HeCd laser (excitation at 325 nm, emission collected at 455 nm). This system was able to detect horse heart myoglobin in the low attomole (injected mass) range. The complexity in fabricating postcolumn reaction systems has to date prevented their commercialization.

Abler et al. [30] developed a postcolumn affinity detection system for proteins. In this system a fluorescently tagged reporter protein was delivered through a liquid junction between the outlet end of the separation capillary and a reaction capillary. Reagent delivery and mixing was driven by application of a vacuum at the outlet of the reaction capillary. In the application described, separated glycoforms of IgG Fc were complexed with fluorescein-labeled fragment B of protein A. Enhanced fluorescence of the complexes against a background of the tagged reporter protein was detected using an argon ion LIF detector.

MASS SPECTROMETRY

On-line coupling of a capillary electrophoresis system to a mass spectrometer enables molecular weight and structural information to be obtained for separated components [31]. The most widely used interface for introducing biomolecules from the CE separation into a mass spectrometer is the electrospray ionization (ESI) system [32]. With this interface, capillary contents are delivered to an atmospheric pressure ionization system by a liquid sheath flow introduced in a coaxial tube by a microflow pump [33]. The capillary is routed through a metallic needle which is held at an electric potential relative to the mass spectrometer. The needle completes the CE circuit and generates the electrospray. As the electrolyte enters the interface, charged droplets are formed at the capillary tip. Evaporation of solvent reduces the volume of the droplets until ultimately bare gas phase ions are formed which are introduced into the mass analyzer. In some cases, a nebulization gas is added to assist in formation of the electrospray or to reduce corona discharge [34]; this modification has been termed ionspray, and has been used with the sheath liquid delivered via a liquid junction as well as by coaxial flow. The various ESI configurations are illustrated in Figure 3. ESI systems enable the CE to be interfaced with single quadripole or time-of-flight (TOF) mass

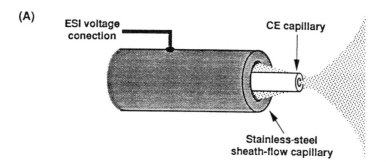

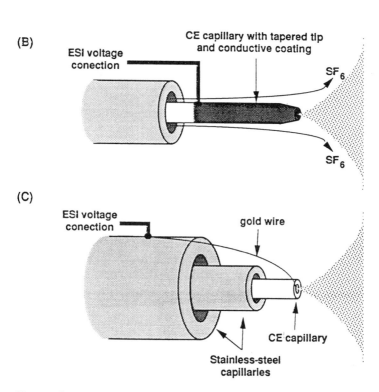

Figure 3 Schematic of CE-ESI-MS interfaces. (A) Coaxial sheath-flow configuration; (B) sheathless interface with tapered and conductive coated CE capillary; (C) sheathless interface using a gold wire electrode. (Reproduced from Ref. 31 with permission.)

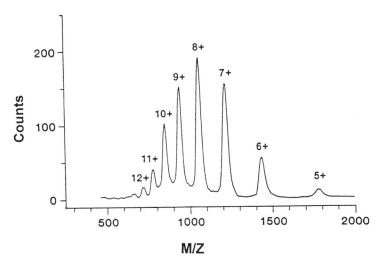

Figure 4 ESI-TOF mass spectrum of ubiquitin. (Reproduced from Ref. 39 with permission.)

spectrometers to obtain molecular-weight information or to triple quadripole or magnetic sector MS/MS systems to generate structural information through collision-induced fragmentation. Ionization of small molecules and peptides typically generates single molecular ions within the 1–4000 mass limit of quadripole MS systems. Proteins, on the other hand, produce multiply-charged ions which appear as a family of peaks in the mass spectrum (Figure 4); because of their multiple charges, these fall within the mass/charge range of the instrument. Liquid junction [35] and coaxial [36] sheath flow designs have also been used to couple CE instruments to fast-atom bombardment (FAB) ionization systems. Performance of these interfaces have been limited by poor sensitivity in the case of the liquid junction method and by the requirement for small i.d. capillaries to minimize laminar-flow mixing induced by the MS vacuum in the case of coaxial flow designs [31].

Although ESI/MS provides the most information-rich detection method for CE, there are limitations to the technique. First, compatibility with mass spectrometry requires the use of volatile buffer systems such as acetic and formic acids or their ammonium salts for low-pH separations or ammonium carbonate for high-pH applications. This limits the selectivity of the CE separation system. Second, the background electrolyte causes discrimination against analyte ions by charge competition in the electrospray, reducing overall sensitivity. This can be minimized by using low concentrations of electrolytes in the CE buffer. Third, the sheath flow liquid also dilutes analyte ions, thereby further reducing sensitivity. This can be minimized by reducing sheath flow rates to 2–5 μl/min and by using sheath liquids composed of low concentrations of electrolyte in organic solvents (e.g., acetic acid in methanol). Differences in the ionic composition of the sheath liquid and the background electrolyte can produce moving-boundary effects which alter resolution and selectivity in the CE separation. This can be minimized by optimizing the composition of the two solutions [37].

The sheath flow can be eliminated entirely by establishing direct electrical contact with the capillary tip. This can be accomplished by etching the outlelt end of the capillary and applying a metallic (e.g., gold or silver) coating [38], or by inserting a gold wire electrode into the capillary outlet [39]. In these systems, referred to as microspray ionization interfaces, small i.d. capillaries are used to increase the efficiency of electrospray formation. A limitation of sheathless interfaces is the requirement for these specially constructed capillaries. Using microspray ionization, nanomolar concentration limits of detection have been achieved for tryptic peptides [40]. Concentration limits of detection can be further decreased to picomolar ranges using a solid-phase on-column concentration method, which also allows elimination of matrix interferences prior to the separation [41]. Sheathless ESI interfaces have also been used with nonvolatile phosphate buffers [42]. Recently a sheathless ESI interface has been coupled to a Fourier-transform ion cycloton resonance mass spectrometer for high-resolution scanning detection of proteins separated by CE [43].

4
Strategies for Reducing Protein-Wall Interactions

Interactions of protein with the inner wall of fused silica capillaries has been the major obstacle to successful application of capillary electrophoresis to protein separations. At pH values above about 2, the weakly acidic silanol groups on the capillary surface become ionized, and the charge density on the wall increases with pH to a maximum of about 10 at which point the silanol groups are fully dissociated. This characteristic of fused silica has two consequences for protein analysis. First, proteins with basic amino acid residues positioned on the protein surface can participate in electrostatic interactions with ionized silanols. Protein adsorption at the capillary wall can result in band broadening, tailing, and, in the case of strong interaction, reduced detector response or complete absence of peaks. Second, changes in the state of the wall during an analysis or from run to run can alter the magnitude of EOF, resulting in changes in analyte migration times and peak areas. Protein adsorption can alter the zeta potential of the capillary wall, changing EOF and degrading reproducibility. Three strategies have been employed to minimize protein-wall interactions: operation at extremes of pH, use of buffer additives, and use of wall-coated capillaries.

OPERATION AT pH EXTREMES

The simplest approach to minimizing protein-wall interaction is to use a buffer pH at which interactions do not occur. Neutral and acidic

37

proteins (pI ≤ 7) will carry net negative charges; under alkaline conditions the capillary surface will also be anionic due to silanol ionization, and protein-wall adsorption will be prevented by coulombic repulsions. The same approach can be used for basic proteins (pI 7–10) if very alkaline buffers are used (e.g., pH 11–12). This approach has been used successfully in some cases [44] but inadequate selectivity and the risk of protein degradation under very alkaline conditions has prevented this from being a general strategy for protein separations. An alternative strategy is operation at very low pH (< 3). Under these conditions, the degree of silanol ionization and the magnitude of EOF are very low. At such slow pH values most proteins will bear a net positive charge, and (in the absence of significant EOF) migrate electrophoretically toward the cathode. Phosphate buffers have proven to be very effective for operation in this pH range because of their high buffering capacity and low UV absorbance. In addition, McCormick [45] has suggested that complexation of phosphate groups with surface silanols contributes to reduced EOF and reduced polypeptide adsorption. However, although use of low-pH conditions has been remarkably successful for separations of complex mixtures of peptides, the approach has been less successful for proteins due to limited selectivity.

Clearly, successful application of capillary electrophoresis to most protein separation problems requires the ability to operate under conditions where protein charge differences are greatest and, preferably, where the proteins are in their native states. Consequently, since the introduction of CE over a decade ago, there have been tremendous efforts to develop conditions for protein analysis at physiological pH in which protein adsorption is minimized and EOF is either minimized or stably controlled.

USE OF BUFFER ADDITIVES

The use of bare fused silica capillaries has advantages in simplicity, low cost, and good capillary lifetime. For this reason, many investigators have searched for buffer components and buffer additives which would permit protein separations to be acheived with uncoated capillaries.

Green and Jorgenson [46] proposed the use of high concentrations of monovalent alkali metal salts to reduce protein adsorption, and determined the order of effectiveness to be $Cs^+ > K^+ > Na^+ > Li^+$. The counter ion had little effect on adsorption but in many cases contributed unacceptably high UV absorbance. Chen et al. [47] applied this approach to the separation of milk proteins. Although effective in reducing protein-wall interactions, use of high-ionic-strength electrolytes generates excessive joule heat. This can be circumvented by use of capillaries with small internal diameter or by operation at lower field strengths, approaches which compromise detection sensitivity and analysis time.

A variety of amine-containing organic bases have been used as additives to reduce protein adsorption (see Table 1). Alkyl amines added to the electrophoresis buffer under conditions where they are protonated will participate in electrostatic interactions with silanols on the capillary wall, stabilizing the state of the double layer and the level of EOF [48,49]. Protein-wall interactions are reduced, resulting in improved peak shape and separation efficiencies. At very low pH, alkylamine additives reverse the direction of EOF toward the anode [50,51]. Use of amine additives increases operating current, and protein-additive interactions may affect the separation. The latter phenomenon may be advantageous in some cases; Landers et al. [52,53] have demonstrated that use of alkyl diamines and bis-quaternary ammonium alkanes provided improved separation of protein glycoforms, which was suggested to arise in part from additive-protein interactions.

The use of zwitterions as buffer additives for capillary electrophoresis was first proposed by Bushey and Jorgenson [54]. Zwitterions should reduce protein-wall interactions without increasing the conductivity of the electrophoresis buffer. Ideally, the additive should be zwitterionic over a wide pH range, have good solubility to enable use at high concentrations, produce a stable electroosmotic flow, and exhibit low UV absorbance. These authors demonstrated that zwitterions such as betaine and sarcosine could be used to improve protein separations, although best results required use of these zwitterions in combination with salts. Since this work was published, a wide variety of zwitterionic compounds have been used for capillary electrophoretic separations of proteins (see Table 1).

Table 1 Additives Used in CZE

Additive	Application	Reference
Betaine	Basic proteins	54
Cadaverine	Acidic and basic proteins	59
Ethylene glycol (20%)	Acidic and basic proteins, serum proteins	60
Cationic and zwitterionic fluorosurfactants	Acid and basic proteins	55–58, 61
2-(N-Cyclohexylamino) ethanesulphonic acid (CHES)	Insulins	62
N,N-bis(2-Hydroxyethyl)-2-aminoethanesulphonic acid (BES)	Insulins	62
3-[(1,1-Dimethyl-2-hydroxyethyl)amino]-2-hydroxypropanesulphonic acid (AMPSO)	Insulins	62
3-(Cyclohexylamino)-2-hydroxy-1-propanesulphonic acid (CAPSO)	Insulins	62
N-Alkyl-N,N-dimethylammonio-1-propane sulfonic acid	Basic proteins	63
1,4-Diaminobutane	Protein glycoforms	48, 52, 64
Spermine,spermidine	Protein glycoforms	65
Triethylamine	Basic proteins	49, 51, 66
Triethanolamine	Basic proteins	49, 51
Galactosamine	Basic proteins	50
Glucosamine	Basic proteins	50
Trimethylammonium propylsulfonate (TMAPS), Trimethylammonium chloride (TMAC)	Monoclonal antibody	67
Trimethylammonium propylsulfonate (TMAPS)	Acidic, neutral, and basic proteins	68
(Trimethyl)ammonium butylsulfonate (TMABS)	Acidic, neutral, and basic proteins	68
2-Hydroxyl-3-trimethyl-ammonium propylsulfonate (HTMAPS)	Acidic, neutral, and basic proteins	68

Table 1   Continued

Additive	Application	Reference
3-(Dimethyldodecylammonio) propanesulfonate	Basic proteins	69
Cetyltrimethylammonium bromide	Acidic and basic proteins	70, 71
Chitosan	Basic proteins	72
Amino acids	Basic proteins	49
Hexamethonium bromide, hexamethonium chloride	Protein glycoforms	53
Decamethonium bromide	Protein glycoforms	53
Polydimethyldially- lammonium chloride	Basic proteins	73, 74
Phytic acid	Acidic and basic proteins	75–77
Ethylenediamine	Basic proteins	48
1,3-Diaminopropane	Basic proteins	48
N,N-Diethylethanolamine	Basic proteins	49
N-Ethyldiethylamine	Basic proteins	49
Morpholine, tetraazomacrocycles	Basic proteins	78

The use of fluorosurfactant additives has been proposed by Emmer et al. (55–58). Both anionic and cationic surfactants or binary mixtures of the two were employed. By varying the proportion of the two surfactants in mixtures, EOF could be varied from strongly positive (normal flow) to strongly negative (reverse flow). Simultaneous separation of acidic and basic proteins was possible with this system.

CAPILLARY COATINGS AND OTHER SURFACE MODIFICATIONS

Surface modifications to the capillary wall may be grouped into two categories, covalent and dynamic coatings. Covalent coatings are attached to the wall through a chemical bond, whereas dynamic coatings physically interact with the silica or a first layer deposited over the silica. Use of dynamic coatings generally requires the incorporation of the wall-interacting compound in the run or conditioning buffer.

Capillary coatings operate by physically blocking the access of the sample molecules to the capillary wall. Coatings and other inner wall modifications are an area of intense research in CE. Much controversy has been generated over which approach is more convenient and produces a more stable modification, thus yielding better and more reproducible results. Perhaps a missing ingredient in the discussion of coatings is the reproducibility and effectiveness of available protocols, e.g., methods fully optimized and standardized for particular coating procedures.

Coatings usually consist of one or two layers. One-layer coatings use a compound that can be reacted with the capillary wall to mask the charges typical of fused silica, and/or to change the characteristics (sometimes dramatically) of the capillary surface. In two-layered coatings the first layer (e.g., bifunctional moieties, hydrophobic layers) is used to anchor the second layer (often polymeric), which typically defines the characteristics of the surface. The main differences among coatings are the type of first layer and/or type of anchorage bond used, and the composition of the second layer. To be useful for routine analysis, a coating must be stable for prolonged periods of time under typical analysis conditions. Loss of coating and changes due to interaction with sample or other buffer components leads to degradation of efficiency and poor reproducibility.

The diversity of chemistries described in the literature reflects the continuous search for more stable coatings, and perhaps the inadequacy of any single approach to provide satisfactory results for all applications. To the CE practitioners who prepare their own capillaries, a particular problem arises because of the high number of nonoptimized coating "recipies" that are continuously being published, and because once optimized, the originally explicit chemistry may be lost behind a commercial trade name.

Coating comparisons do not always lead the user into better chemistries since optimization is achieved for only one of the methods compared. Hjertén [89] noted that even the same coating process does not produce the same quality coating from column to column, and this may reflect variations in the quality of the fused-silica capillary. Thus, he suggested that capillary pretreatment and coupling chemistries may have to be optimized for different sources of capillary or even for different batches of silica.

Coating chemistries have been extensively reviewed (79–81) and a description of each coating is beyond the scope of this book. Nevertheless, some historically important and new developments can be addressed.

One-Layer Wall Modifications

Silanized capillaries with carbon moieties ranging from 1 to more than 18 have been used to render the capillary wall more hydrophobic. In most cases buffer additives are used to avoid sample adsorption and/or to control EOF.

Polybrene is a positively charged polymer that adsorbs tenaciously to the silica wall and has been used to reverse the charge of the capillary wall [82], thus allowing the analysis of basic proteins at pH values below their pI (Figure 1). This coating has also been employed for the analysis of small molecules (e.g., inorganic ions). A similar approach was employed by Towns and Regnier [83] who used poly(ethylenimine) adsorbed to the capillary wall to form a cross-linked cationic polymeric coating. Like polybrene, the cationic nature of the PEI coating produced a reversal of EOF and eliminated adsorption of proteins at pH values below their pI. Cifuentes et al. [84] observed that both the magnitude of reversed EOF and separation selectivity varied with buffer pH. Addition of PEI to the running buffer significantly affected selectivity, presumably due to electrostatic interactions of proteins with the basic polymer additive.

Several compounds have been used to cover the capillary wall by filling the column with a solution containing the material, flushing it out (usually with an inert gas), leaving a thin layer of the compound deposited on the wall when the solvent evaporates. Typically, the solute used is not easily dissolved by the buffer system used for electrophoresis. Polyvinylalcohol [85,86], poly(ethyleneoxide) [87], and cellulose acetate [88] coatings are examples employing this technique.

Two-Layer Wall Modifications

Polyacrylamide (PA) Coatings

One of the most popular wall coverage materials is polyacrylamide, and in this section we describe techniques that maintain the second

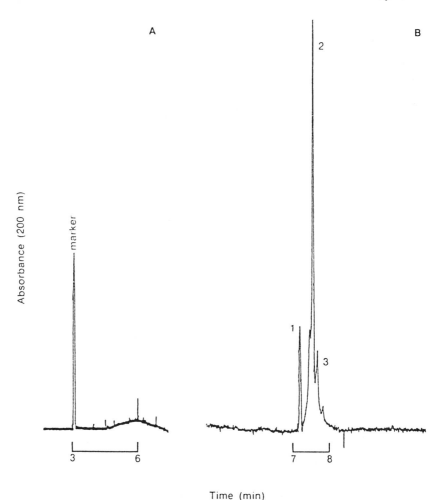

Figure 1 (A) Application of isoenzymic forms of LDH in an uncoated capillary. Conditions: Field strength, 400 V/cm; current, 4 μA; buffer, 5 mM sodium phosphate, pH 7.0; capillary length, 90 cm (68 cm effective length); capillary diameter, 50 μm; separation temperature, 30°C; polarity normal. (B) Separation of LDH isoenzymes under conditions of capillary charge reversal. Peaks 1, 2, 3 are isoenzymes of pI 8.3, 8.4, and 8.55, respectively. Conditions same as (A) except that polarity is reversed. Horizontal scale in minutes. (Reproduced from Ref. 82 with permission.)

layer (polyacrylamide) constant, but differ in the way it is anchored to the wall.

Hjertén [89] introduced a new method of coating capillaries shortly after successfully adapting the IEF process to the capillary format. In his approach, linear hydrophilic polymers are chemically attached to the wall via a bifunctional group (siloxane chemistry) which also provides a vinyl group used to anchor the polyacrylamide chains polymerized in situ. Hjertén not only published a procedure to synthesize coated capillaries, but also described the theoretical basis for elimination of nonspecific adsorption and EOF. His discovery that EOF is reduced as the viscosity near the ionic double layer (Stern layer) is increased has been exploited using other ligands and bonded chemistries, and is also behind the rationale of some dynamic coatings. Typically, a well-coated linear PA capillary exhibits over a 50-fold EOF reduction as compared with uncoated columns.

Polyacrylamide has been attached to a vinyl group directly anchored to a chlorinated silica by reaction with a Grignard reagent [90]. This reaction produced a silicon–carbon bond. The vinyl group was then reacted with acrylamide. Through this procedure electroosmosis is virtually eliminated. The silicon–carbon bond provided high stability, even when exposed to pH 10.5 for 5 days.

Linear polyacrylamide can be cross-linked after attachment to a capillary first treated with 7-oct-1-enyltrimethoxysilane [91]. This silane is known to cross-link extensively and, according to the authors, renders a more stable first layer to which the polymer is anchored.

Schmalzing et al. [92] developed a coating that used vinyl siloxanediol to anchor linear or cross-linked acrylamide. The hydroxyl groups of the siloxanediol were used as cross-linkers in the first layer. Acrylamide was reacted with the vinyl group of the deposited siloxanediol, then the resulting polyacrylamide chains were cross-linked by filling the column with a 37% solution of formaldehyde (pH 10, adjusted with NaOH) and incubating for 3 h. Linear PA reduced EOF more efficiently than cross-linked PA, but the latter showed a lower degree of EOF change as a function of operating time at a pH of 8.8.

Other Polymeric Coatings

Polymeric coatings can also be used to produce capillaries with a desired level of EOF [93]. In one case, the column was covered with a mixture monomers of acrylamide and 2-acrylamido-2-methyl-1-propanesulfonic acid (AMPSA). The tube was treated with a bifunctional silane before polymerization of the final layer. The resulting capillaries exhibited a remarkably constant EOF value over a pH range of 2–9 as compared with bare silica capillaries. Xu et al. [94] polymerized vinylpyrrolidone (VP) and vinylimidazole (VI) onto capillaries derivatized with methacryloxypropyltrimethoxysilane to form polyVP, polyVI, and copoly (VP + VI) coatings. These coatings did not completely mask the silica surface, as indicated by low to moderate EOF toward the cathode, although positive charges introduced by VI prevented adsorption of basic proteins. Efficiency and reproducibility were satisfactory at pH 5–6, but the coatings were unstable at pH 7. Liu et al. [95] prepared covalently anchored epoxy resin coatings cross-linked with an aromatic amine which exhibited constant EOF over neutral-to-alkaline pH ranges and good stability up to pH 12.

Srinivasan et al. [96] developed a method for attachment of preformed polymers to a silanized capillary surface using a free-radical process which covalently coupled the polymer to the silane anchor and cross-linked the polymer chains. Neutral polymers (PVP, PEO) produced coatings with low EOF and stability to alkaline conditions, while a cationic polymer prepared using a quaternary ammonium acrylamide monomer produced a coating with reversed EOF.

Some natural polymers exhibit very high hydrolytic stability even at extreme pH, and for that reason Hjertén and Kubo [97] attached dextrans and cellulose derivatives, and Mechref and El Rassi [98] attached high-molecular-weight dextrans to the capillary wall. The method described by El Rassi is composed of three steps: first the silica was treated with an oligomeric epoxysilane by incubating at 96°C; then the dextran was anchored to the first layer in the presence of boron trifluoride; finally, the dextran was cross-linked with diepoxypolyethylene glycol in the presence of boron trifluoride. Capillaries prepared by the above method were exposed to various harsh conditions (e.g., dilute NaOH) to test for stability, and a group of selected proteins analyzed to demonstrate the performance of the columns. EOF reduction was found to be affected by the size of the dextran used, with lowest EOF achieved with a 150 kD dextran (7–10 times EOF reduction as compared with uncoated capillaries). Model basic proteins analyzed at pH 5 yielded up to 750,000 plates/meter (Figure 2). Acidic proteins showed a decreased number of plates/meter (111,000).

Cellulose derivatives have also been attached to the capillary wall using several procedures [91,99]. Cellulose coated capillaries were found to withstand 30 days of exposure to 10 mM NaOH (pH 12) [88].

A problem encountered in our laboratory working with natural polymers is that they tend to provide a less efficient wall coverage as

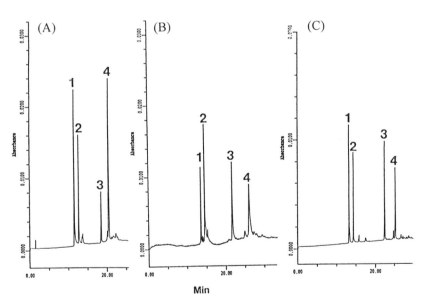

Figure 2 Typical electropherograms of four standard basic proteins obtained with capillaries coated with 45 kDa dextran (A), 71 kDa dextran (B), and 150 kDa dextran (C). Running electrolyte: 100 mM phosphate, pH 5.0; pressure injection 1 s; applied voltage, 15 kV. Samples: 1, lysozyme; 2, cytochrome C; 3, ribonuclease A; 4, α-chymotrypsinogen A. (Reproduced from Ref. 98 with permission.)

Table 2 Commercially Available Coated Capillaries

Coating type	Source	Characteristic	Applications
Linear Polyacrylamide	Bio-Rad Laboratories	Linear; hydrophilic	Polypeptides, peptides, nucleic acids, small molecules
Proprietary	Beckman Instruments	Hydrophilic	Polypeptides, nucleic acids
CElect-P	Supelco, Inc	Hydrophilic	Polypeptides
CElect-H	Supelco, Inc	Weakly hydro-phobic (C1)	cIEF
CElect-H1	Supelco, Inc	Moderately hydro-phobic (C8)	
CElect-H2	Supelco, Inc	Highly hydro-phobic (C18)	
Micro-Coat	Applied Biosystems	Dynamic, cationic	Polypeptides
CE-100-C18	Isco, Inc	Hydrophobic (for use with dynamic coatings)	Polypeptides
CE-200-Glycerol	Isco, Inc		Polypeptides
CE-300-Sulfonic	Isco, Inc		Nucleotides
DB1	J & W Scientific	Hydrophobic	cIEF
Proprietary	Dionex		Nucleic acids
MicroSolv CE	MicroSolveCE	Deactivated	Polypeptides
PEG Coated	MicroSolveCE	Weakly hydrophilic	Polypeptides

judged by the presence of higher EOF values compared to compounds polymerized in situ. One possible explanation is steric hindrance of the attached molecules with the incoming ones to unoccupied available patches on the column surface. A good coverage would require a perfect ordering side by side of such bulky molecules.

Chiari et al. (100) developed N-acryloylaminoethoxyethanol (AAEE) monomers which can be used to form a polymeric coating. The authors demonstrated that this compound exhibits a much higher hydrolytic stability than acrylamide, and is also more hydrophilic. Coatings prepared with this material show improved performance and higher stability (the authors' laboratory, unpublished data).

Surfactant-Coated Capillaries

Towns and Regnier [101] derivatized the capillary surface with octadecyltrichlorosilane to form a covalently bonded hydrophobic phase. When an aqueous solution of a nonionic oxyethlene-based surfactant (e.g., Brij 35) at a concentration above its critical micelle concentration was introduced into the capillary, the hydrophobic portion of the surfactant formed an adsorbed layer on top of the covalently bonded (silane) hydrophobic layer. The polar head group of the surfactant was exposed to the aqueous solution filling the column, forming a hydrophillic layer that masked the hydrophobic layer and the unreacted silanol groups, thus effectively reducing protein interaction with the surface of the capillary.

A list of commercially available coated capillaries, their characteristics, and suggested applications are presented in Table 2.

5

Capillary Zone Electrophoresis

Capillary zone electrophoresis (CZE) offers several advantages in comparison to other CE separation modes. In most cases, CZE is simple and straightforward to perform: a single buffer is used throughout the capillary and electrode vessels, and sample is introduced as a zone or plug at one end. Capillary preparation often involves only filling the capillary with the separation buffer, although uncoated capillaries generally require prior washing or conditioning steps. The technique is inexpensive in comparison to capillary gel electrophoresis or isoelectric focusing, since low-cost buffers and salts are used. Separation selectivity can be easily varied by manipulation of buffer pH or use of additives. For these reasons, CZE is the most widely used CE mode for protein separations.

DEVELOPING A CZE METHOD

The development of a CZE separation is typically a three-step process: determination of the optimum buffer composition, selection of the appropriate capillary type, determination of high-voltage power supply parameters.

Buffer Selection

Selection of the analysis buffer will be dictated by the pH value required to achieve satisfactory analyte mobility and resolution. To achieve the first goal, the pH of the buffer should be at least one pH unit above or below the isoelectric point of the analyte protein; at pH values closer to the protein pI, low mobility will result in long analysis times, peak broadening, and increased risk of protein-wall interactions. To achieve satisfactory resolution, the buffer pH should provide satisfactory differences in the mass-to-charge ratios of analyte proteins so that they exhibit significant differences in mobility. In many applications, a sample may contain several proteins with such a wide range of isoelectric points that some sample components may take a very long time to reach the detection point or may (in the absence of EOF) migrate in the opposite direction. In such cases two strategies can be used: the separation may be performed at extremes of pH where all proteins possess net positive charge (e.g., pH 2–2.5) or net negative charge (e.g., pH 10–11), or the separation may be run twice at different polarities. When operating under alkaline conditions with uncoated capillaries, the high level of EOF will sweep acidic proteins past the detection point, allowing separations to be achieved for a mixture of acidic and basic proteins using positive-to-negative polarities. However, as discussed above, adsorption of basic proteins to the silica surface may alter the level of EOF and lead to poor reproducibility.

The buffer should have good buffering capacity at the selected pH to provide good reproducibility; high buffering capacity is advantageous since lower ionic strength may be used to minimize Joule heat. Low-conductivity buffers are also desirable to minimize thermal effects on protein structure.

Buffers with low absorbance in the UV region must be used to achieve low background noise and satisfactory detection limits. This is not a limitation in selecting buffers for conventional gel electrophoresis since proteins are stained following the separation. Because CE employs on-tube absorbance detection, many of the buffers commonly used in gel electrophoresis are unsuitable for CE since they produce unacceptably high background at 200 nm. In some cases use of slightly longer wavelengths (e.g., 210–220 nm) reduces background noise significantly without a serious decrease in analyte signal so that satisfactory detection limits can be achieved.

A list of buffer systems is provided in Table 1 with characteristics relevant to their use for CZE separations. Current and absorbance values were determined under typical CZE analysis conditions (24

Table 1 CZE Buffer Performance Characteristics

Buffer Ion	Counter Ion	Molarity (mM)	pH	pK of Buffer Ion	Current	Absorbance at 200 nm (mAU)	Absorbance at 220 nm (mAU)
Phosphate	Sodium	100	2.0	2.12	66	3	3
Phosphate	Tris	50	2.0	2.12	37	3	0
Citrate	Citrate	100	3.0	3.06	25	807	47
Formate	Tris	50	4.0	3.75	24	342	70
Acetate	Tris	50	5.0	4.75	21	612	46
MES	Sodium	100	6.0	6.15	17	2458	273
MES	Tris	100	6.0	6.15	16	2356	134
MOPS		100	7.0	7.20	27	2564	344
Taurine	Sodium	100	8.0	8.95	9	504	28
Taurine	Sodium	200	8.0	8.95	17	1008	28
Phosphate	Tris	100	8.0	8.30	35	1000	0
Bicine		100	8.0	8.35	20	3260	2368
HEPES	Tris	100	8.0	8.00	27	2752	1105
Boric acid	Sodium	100	8.5	9.24	18	-0.15	-0.34
CHES		100	9.0	9.50	17	2025	270
Phosphate	Tris	200	9.0	8.30	19	2363	41
Beta alanine	Sodium	100	10.0	10.19	20	2272	160
CAPS	Sodium	100	10.0	10.40	13	1908	281
Glycine	TEA	40	10.2	9.60	20	1612	192
CAPS		50	11.0	10.40	25	2429	351

Current measured using a 24 cm × 50 μm capillary at 10,000 V.
Absorbance measured against water using a 24 cm × 50 mm capillary.

cm × 50 μm i.d. capillary operated at 10 kV). The properties of these buffers illustrate the difficulties in selecting a satisfactory CZE buffer for a desired pH range. The zwitterionic buffers MES, CAPS, CHES, β-alanine, and bicine are desirable for their low conductivity, but exhibit high absorbance in the low UV region. Taurine buffers offer satisfactory buffering capacity, relatively low UV absorbance, and low conductivity in the pH 8–8.5 range. Phosphate buffers are desirable for low UV absorbance and low conductivity in the pH 2–3 range, but exhibit high conductivity at pH 6–7. Borate buffers are excellent choices for the pH 8–9 range because of their low UV absorbance, good buffering capacity, and low conductivity.

It is often necessary to add additional components to the analysis buffer to achieve a satisfactory separation. Buffer additives such as neutral salts, zwitterions, alkyl amines, and neutral polymers may be used to reduce or control EOF and protein adsorption, and these additives are discussed in Chapter 4. Other additives may be used to solubilize proteins or to reduce protein aggregation. These may be required when analyzing transmembrane proteins, structural proteins, or large proteins with significant hydrophobic surface area. Solubilizing agents may also be necessary when separating core fragments generated by cyanogen bromide cleavage. Chaotropes such as guanidium chloride or urea are sometimes used in chromatographic mobile phases to improve protein solubility and recovery. Guanidinium salts are impractical in capillary electrophoresis because of excessive Joule heating. Urea can be used in concentrations up to 8 *M*. However, high concentrations of urea introduce high background absorbance levels, and it may be necessary to increase the detection wavelength to 215–220 nm to obtain acceptable baseline noise. A more acceptable approach is the use of 7–8 *M* urea in the sample to solubilize proteins and use of a lower concentration (4–6 *M*) in the analysis buffer to maintain protein solubility. Use of urea in CE systems requires great care to achieve good instrument performance. When used at such high concentrations, urea easily contaminates components of the CE, and current leakage and arcing are common problems when using this additive. Attention to system cleanliness is extremely important; the high voltage electrodes and capillary housing should be inspected daily, and if urea crystallization is observed, the contaminated surfaces should be cleaned with deionized water and dried. Exposure of buffer-containing capillaries to the air will result in urea crystallization and capillary plugging. After use, capillaries should be immediately purged with water or stored with the tips immersed in buffer. Because urea is very water-soluble, plugged capillaries can often be recovered by immersing the capillary tips in warm (70°C) deionized water for 30 min, or in water at ambient temperature overnight. If this treatment is unsuccessful, the capillary tips can be immersed in a sonic bath containing deionized water for several minutes.

Surfactants can also be added to improve protein solubility. Surfactants should be selected for their solubilizing power and low UV background. Ionic surfactants may contribute to Joule heating, so zwitterionic or neutral surfactants may be preferable. Brij 35 and Triton X-100 have been used with good success at concentrations up to 1% w/v. The reduced (hydrogenated) form of Triton X-100 must be used to achieve low baseline noise, and it exhibits similar surfactant properties to unreduced Triton X-100. A number of the common surfactants used for solublizing proteins are listed in Table 2 with characteristics relevant to their use in CZE of proteins.

Organic solvents may used as additives for hydrophobic proteins. Glycerol may be added in concentrations up to 20%. This additive will increase the buffer viscosity, and may require adjustments to capillary purge volumes and injection parameters when using displacement injection (pressure, vacuum, or gravity). Alcohols such as methanol, ethanol, propanol, or isopropanol by be used as buffer modifiers. Like glycerol, they will change buffer viscosity. In addition, short-chain alcohols have significant vapor pressure and will evaporate if buffer reservoirs are not tightly capped. Solvent evaporation can result in variability in injection volume due to changing viscosity, and to variable migration times due to changing buffer conductivity.

Additives may also be used to maintain an analyte protein in a desired charge or conformational state. The migration behavior of metal-binding proteins may vary depending on the amount of bound metal ions, and CZE may be used to determine the binding properties of a protein by varying the level of ligand in the analysis buffer.

Table 2 CZE Surfactant Performance Characteristics

Surfactant	Molecular Weight	Aggregation Number	CMC (mM)	Current	Absorbance at 200 nm (mAU)	Absorbance at 200 nm (mAU)
Brij 35	1200	40	0.05–0.1	5.19	5.2	1.5
CHAPS	615	4–14	6–10	0.40	102.5	3.7
Cholate, Sodium	431	2	9–15	11.63	17.6	0.4
Deoxycholate, Sodium	415	3–12	2–6	12.81	17.3	0.2
Dodecylsulfate, Sodium	289	62	7–10	11.20	−0.7	−0.6
Octylglucoside	292	84	20–25	0.61	−0.4	−0.6
Taurodeoxycholate, Sodium	522	6	1–4	9.23	109.4	4.2
Triton X-100, Reduced	631	—	0.25	2.20	13.8	2.5
Tween 20	1228	16.7	0.059	0.38	5.8	2

Current and absorbance of 1% aqueous solution measured using a 24 cm × 50 μm capillary at 10,000 V

Similarly, CZE can be used to determine enzyme binding constants for charged ligands such as substrates, inhibitors, and cofactors by observing the effect of ligand concentration on migration; this technique has been termed affinity capillary electrophoresis (ACE) and is described in detail in the Applications section. If the product of an enzymatic reaction is both charged and UV-active and can be resolved from the enzyme, CZE can be used to assay enzyme activity. This technique, termed enzyme-mediated microassay (EMMA) is also described in the Applications section.

Buffers should be filtered and degassed prior to use. Degassing can be accomplished by vacuum filtration or by centrifugation for 2–5 minutes in a microcentrifuge. It is advisable to store buffers at 4°C to prevent microbial growth, since even slight contamination by microorganisms can cause capillary plugging and detector interference.

Capillary Selection

Fused silica capillaries are almost universally used in capillary electrophoresis. The inside diameter of fused silica capillaries varies from 20 to 200 μm and outside diameter varies from 150 to 360 μm. The capillary is coated with polyimide to provide mechanical stability, and this opaque polymeric coating must be removed at the detection point. This is most easily accomplished by heating the detection segment in a flame, by dripping hot concentrated sulfuric acid on the coating, or by scraping with a sharp blade such as a scalpel. This detection "window" is very fragile, and once the window is created the capillary must be handled with great care to prevent breakage. Capillaries coated with a UV-transparent cladding are commercially available, and these can be used directly without removal of the coating. However, the coating is not stable in high concentrations of organic solvents, and is attacked by the fluorocarbon liquid used as capillary coolant in some CE instruments.

Selection of the capillary inner diameter is a tradeoff between resolution, sensitivity, and capacity. Best resolution is achieved by reducing the capillary diameter to maximize heat dissipation. Best sensitivity and capacity is achieved with large internal diameters. A capillary internal diameter of 50 μm is optimal for most applications, but diameters of 75–100 μm may be needed for high sensitivity or for micropreparative applications. However, capillary diameters above 75 μm exhibit poor heat dissipation and may require use of low-conductivity buffers and low field strengths to avoid excessive Joule heating.

Small-bore capillaries with internal diameters of 25 μm or less are sometimes used with high ionic strength buffers because of their better heat dissipation. However, they are subject to plugging, and can increase the risk of protein-wall interactions due to their high surface-to-volume ratios. Capillaries with external diameters of 360 μm are the most widely used, although smaller o.d. capillaries are commercially available. Capillaries with reduced outer diameter and large i.d. (e.g., 50 μm i.d. × 150 μm o.d.) exhibit greater fragility due to reduced wall thickness, and may fracture when subjected to high voltage. Short capillaries with large internal diameters are more susceptible to siphoning which may compromise reproducibility and introduce laminar-flow band broadening.

Selection of capillary length is dictated by the type of capillary used and the required resolution. When using coated capillaries with insignificant EOF, separations can be achieved with relatively short capillaries of 20–30 cm effective (inlet to detection point) length. When using uncoated capillaries under conditions where there is appreciable EOF, longer lengths of 50 cm or greater may be needed to achieve a separation, particularly for basic proteins which are migrating toward the detector under the combined forces of EOF and electrophoresis. Capillary coatings, described in detail in Chapter 4, can be neutral to suppress protein adsorption or charged to reverse the direction of EOF and reduce adsorption of basic proteins.

High-Voltage Parameters

High-voltage parameters include mode of operation, field strength, and polarity. Commercial CE systems can be operated in constant voltage, constant current (in some instruments), and constant power mode. The great majority of protein separations reported in the literature have been performed in constant voltage mode. Constant current operation may be desirable in systems without adequate temperature control; with well-designed liquid or forced-air temperature control systems, constant voltage operation provides good reproducibility. Constant power mode allows separation time to be minimized without excess heat generation, although there are very few reports of separations achieved using constant power operation. Selection of field strength is a tradeoff between resolution and analysis time. Operation at high field strength reduces analysis time but increases band broadening due to thermal effects; operation at low field strength reduces heating but increases analysis time and band broadening due to diffusion. The latter effect is probably small for proteins because of their low diffusion coefficients. In our experience, operation at 400–600 V/cm provides optimal separations in terms of speed and resolution. Selection of high voltage polarity depends on the sample composition and capillary type. When using uncoated capillaries under neutral to alkaline conditions, positive (inlet side) to negative (detector side) polarity is employed so that EOF will carry all samples toward the detection point. This is often termed "normal" polarity. When uncoated capillaries or capillaries coated with a neutral material are used with an acidic analysis buffer (pH 2–3), positive-to-negative polarity should be used since all proteins will carry a net positive charge. When coated capillaries are used between pH values of 3–9, polarity selection will depend on the isoelectric points of the proteins of interest. When using capillaries coated with cationic functionalities, the direction of EOF will be reversed, requiring operation with negative-to-positive ("reversed") polarity.

SAMPLE PRECONCENTRATION TECHNIQUES

Sample preconcentration techniques are used for two purposes: (1) to increase concentration in order to achieve detection; (2) to eliminate disturbances of the electrophoretic system when the conductivity of the sample is significantly different from the background electrolyte (during hydraulic or electrokinetic sample introduction).

Preconcentration to Improve Detection

The primary advantage of performing electrophoresis in narrow bore capillaries is higher resolution. The main problem associated with the use of capillaries is detection. Since the most widely used methods of detection in CE are based on optical absorptivity, the signal generated depends on the path length the light has to travel (along with the concentration and absorption coefficient of the sample). In CE the optical path is usually only 25 to 100 μm (the i.d. of the capillary for on-line detection), so the signal generated by the detector is weak. For this reason, CE requires relatively high initial sample concentration.

For sample components present at low concentartions, several strategies have been described to preconcentrate. These techniques include zone sharpening [102,103], on-line packed columns [104], and transient cITP [105,106].

Principles of Concentration

Most concentration procedures utilize some form of transport phenomena to achieve concentration. The principle is to mobilize (transport) sample molecules in space, slowing the velocity of the leading particles and/or accelerating the trailing ones, thus allowing the trailing molecules to catch up leading molecules. Differential transport can be achieved by using temporary physical barriers and/or uneven transport force distribution. Chromatographic methods can also be used to trap the molecules present in low concentration but large volumes, and then desorb them with a reduced volume of solvent.

The principle of using physical barriers to concentrate is better illustrated by using a classical technique such as ultrafiltration as an example. In this method, the leading particles are stopped by a dialysis membrane, allowing the molecules in solution to eventually reach the same physical space at the membrane, thus increasing concentration. Since the membrane is semipermeable, desalting occurs concomitantly with concentration.

Electrophoretic techniques can be used to concentrate by manipulating the voltage distribution along the capillary axis, or by changing the analysis conditions [105] (e.g., pH) after the concentration step has been achieved. Two concentration methods based on voltage distribution are described below: zone sharpening and isotachophoresis.

Preconcentration Using Physical Barriers

For low volumes of solutes, Hjertén et al. [107] developed physical barriers (gels, dialysis membranes) directly attached to one end of the capillary. In this approach, the capillary is filled with the sample diluted in the leading buffer (ITP is later used as a "mobilization" step) and the tip of the column is pressed against a polymerized acrylamide gel (T = 40%, C = 3%) containing buffer, and then drawn slowly off the gel. This process leaves a gel plug equal to the depth to which the capillary is pushed into the gel. An electric field is applied with the polarity set so that migration of the proteins proceeds in the direction of the gel. Since the pores of the gel are very small, the proteins accumulate on its surface. Once the concentration process is completed, a mobilization step (e.g., ITP) using a terminating buffer for a short period of time is used to avoid peak broadening. After mobilization, the separation proceeds (as CZE) by replacing the trailing buffer for a vessel containing leading electrolyte (Figure 1). When the physical barrier is a dialysis membrane, the proteins

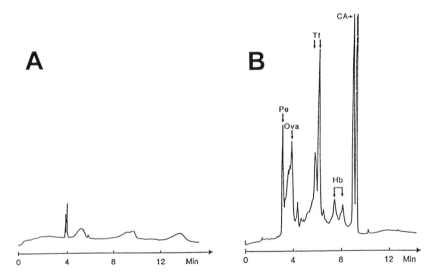

Figure 1 High-performance capillary zone electrophoresis of model proteins. (A) Prior to concentration (concentration of each protein: 20 µg/ml); (B) following concentration toward a small-pore polyacrylamide gel and a short mobilization by displacement electrophoresis. The width of the applied sample zone was in (A) 3–4 mm and in (B) 140 mm (= the length of the capillary). Following concentration the zone width in (B) was about 0.2 mm. (Reproduced from Ref. 112 with permission.)

may be detached from the membrane using a hydrodynamic force (e.g., by raising the electrode vessel closest to the membrane).

Preconcentration by Electrical Force Distribution and/or Mobility Manipulation

The first and simplest method of on-line sample preconcentration is zone sharpening by stacking [108]. In this procedure the sample is dissolved in a buffer of lower conductivity (most commonly diluted run buffer) than the run buffer. Upon applying an electric field, the sample components become concentrated at the interface of the sample and run buffers. The key element of zone sharpening is the lower conductivity of the sample solution as compared to the run buffer. Maximum concentration would, theoretically, be achieved in complete absence of any salt ions (including buffer ions) in the sample solution. Under these conditions, the sample molecules must carry all the electric current in the portion of the system occupied by the sample solution. Since (most) sample molecules have relatively low mobilities, a high voltage drop occurs in this region, and the sample ions are accelerated. The sample-buffer interface posseses buffer ions (higher conductivity) and thus lower voltage. When the sample molecules reach this area, they are deccelerated and the trailing ions, still moving fast toward this zone, eventually catch up with them, and thus the sample becomes concentrated. When the differential conductivity is removed (by replacing the sample solution with run buffer in electrophoretic injection, or by ion diffusion during hydraulic injection) the voltage is again distributed evenly along the capillary, and electrophoresis proceeds normally. Besides increasing sample concentration, zone sharpening also improves resolution by narrowing the starting zones.

Zone sharpening is less effective as the concentration of buffer ions (or any other competing ions) in the sample is increased. For this reason, sample desalting is often necessary in CE. Special desalting methods are required when the characteristics of the sample (e.g., very low volumes) do not allow the use of classical techniques (e.g., ultrafiltration, dialysis).

Hjertén et al. [109] described an off-line method for desalting and concentrating microliter volumes of protein samples using small-pore polyacrylamide gels. The gels can be prepared in several formats:

tube gels with a fused silica "piston," small gel-filled test tubes with wells, or gel-filled Petri dishes containing wells. The tube gel format is used as a syringe, allowing an individual sample to be introduced by drawing the piston. The other two formats are useful for processing multiple samples. Gels may be cast in a low-ionic strength buffer (typically 10-fold dilution of the electrophoresis buffer) at a polyacrylamide concentration providing a pore size appropriate for the application (22%T, 8%C for desalting or 18%T, 5%C for desalting with concentration). The pore size of the gel allows small molecules to diffuse away from the sample, and this was proven to be advantageous when analyzing complex matrices (e.g., amniotic fluid), in which the pattern was greatly simplified (Figure 2).

Preconcentration by Isotachophoresis

A typical setup for isotachophoresis includes a leading electrolyte and a trailing electrolyte with a higher and lower mobility, respectively, than any of the sample components. Another important characteristic of the ITP system is the presence of a counter-ion (of charge opposite to the sample, leading, and trailing electrolytes) that typically possesses buffering capacity. Upon the application of an electric field, the sample components are distributed according to mobility, with the sample ion of highest mobility just behind the leading electrolyte, and the sample ion of lowest mobility just in front of the trailing electrolyte. The voltage gradient is unevenly distributed along the capillary, with the voltage being "concentrated" as the mobility of the ions decrease. Once the system reaches a steady state, the sample components migrate as contiguous zones, since there is no background electrolyte in the sample zones. Following the ITP concentraton step, a CZE separation is used to achieve resolution of contiguous zones.

The uneven distribution of the voltage eliminates diffusion, since an ion that diffuses into a zone of higher mobility encounters a lower voltage, and slows down. If it diffuses into a trailing zone, the voltage is higher, and thus it speeds up. The key phenomena exploited for ITP preconcentration is that the concentration of the sample zones always reaches a value that is determined by the composition of the leading electrolyte [110]. Obviously, the composition of the leading

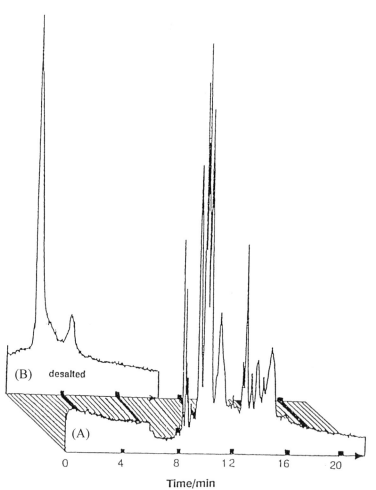

Figure 2 (A) Free zone electrophoresis of nondesalted amniotic fluid. Buffer: 100 mM Tris-acetate, pH 8.6. (B) Free zone electrophoresis of desalted amniotic fluid. Following desalting only two protein peaks appear in the electropherogram, which facilitates clinical diagnostics. (Reproduced from Ref. 109 with permission.)

electrolyte can be manipulated to achieve concentration. It is important to note that often it is necessary to destack the samples before proceeding with the CZE separation. Destacking procedures are also referred as "mobilization" by some authors. The main advantage of

ITP preconcentration compared to zone sharpening by stacking is that it can be used in the presence of salts [106].

Preconcentration by ITP can be performed on-line or in coupled columns. The former uses a transient ITP step and gradually changes into a CZE mode to achieve separation (Figure 3). ITP preconcentration in coupled columns uses one column for ITP concentration and another column for CZE separation. An advantage of transient ITP is the simplicity of instrumentation, since it can be performed in most commercially available CE instruments. Coupled-column ITP-CZE offers a greater potential to concentrate samples, since the injection volume can be increased as compared with transient ITP, where a major portion of the capillary has to be reserved for the CZE stage of the analysis. However, it requires construction of a homemade system.

Foret et al. [106] described general strategies for performing sample preconcentration by ITP. Transient as well as coupled-column ITP were amply illustrated through the use of model proteins. These authors emphasized the need to optimize the preconcentration step, and provided equations to determine optimal preconcentration as well as destacking times, which depend on the composition and the length of capillary originally occupied by the sample. Since the ionic strength of the sample matrix causes variability in migration time, sample desalting was found to increase reproducibility. To produce quantitative results, constant current was used, since under this condition the sample zones move at a constant speed when they pass the detection point. When constant voltage is used, quantitation may suffer from the transient conductivity states of the sample components. Two methods were studied by the authors. The first was a typical ITP setup (trailing electrolyte-sample-leading electrolyte), and after concentration the trailing electrolyte was replaced by leading electrolyte to transform the run into a CZE analysis. Through the use of ITP, the sample (model proteins) introduced into a 50 cm $\times$ 75 μm column (internally coated with polyacrylamide for most analysis) was increased nine-fold (450 nl) compared to a typical CZE injection (50 nl). Likewise, detection limit was decreased almost 10-fold. Migration time reproducibility was about 1% RSD, whereas the area was about 5% RSD. The second method employed fast-moving ions (leading electrolyte) in the sample and the background electrolyte (BE) was a slow-moving ion (trailing electrolyte). In this case the BE was

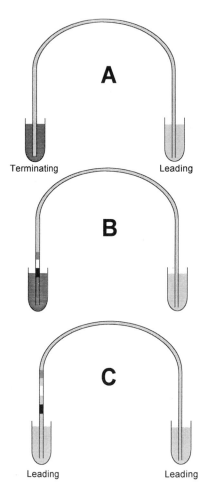

Figure 3   Schematic of on-line transient ITP preconcentration. (A) Initial conditions: the inlet (left) reservoir contains sample dissolved in terminating buffer and the capillary and outlet (right) reservoir contain leading buffer. (B) Preconcentration phase: sample components are resolved isotachophoretically into contiguous zones between the leading and terminating buffers. (C) Separation phase: the inlet reservoir contents are replaced with leading buffer, and sample zones migrate by zone electrophoresis towards the detector.

the one eventually used for CZE, while in typical ITP systems the leading electrolyte becomes the BE after replacing the trailing solution. Upon the application of an electric field, the ions automatically formed characteristic ITP zones, but eventually the concentration of the leading electrolyte (present in quantities determined only by its level in the original sample plug) decreases by diffusion, the electric field eventually becomes homogeneous, and the process is transformed into CZE without the need to replace any of the initial solutions. Coated capillaries were required for the analysis of basic proteins, since samples containing low concentrations of proteins (less than $10^{-8} M$) were lost to adsorption on the capillary wall. Detection limits for the protein mixture after ITP preconcentration were estimated to be around $10^{-9} M$.

A simpler system [111] using peptidelike compounds was employed to study the effect of injection volume, ITP times, and electrolyte composition in the performance of preconcentration by ITP. It was found that if destacking is initiated before completion of preconcentration, the process is not efficient (low signal-to-noise ratio). If destacking is prolonged, resolution is diminished: since both processes occur in the same capillary, prolonged stacking diminishes the separation length of capillary resulting in incomplete separation.

Preconcentration by Conductivity Gradient

An uneven electric field can be obtained by manipulating the conductivity of one or more segments of the electrophoresis system. In one approach [107,112] an electrode vessel was filled with a buffer of very high ionic strength (high conductivity results in low voltage across the reservoir). After concentration the sample was moved into a dialysis tubing attached to the capillary and then back into the capillary. By reversing the polarity of the electric field, dialysis of the sample resulting in a zone sharpening.

Preconcentration by Sample Focusing

Sample focusing [107,112] is very similar to isoelectric focusing (IEF), but does not employ ampholytes (Figure 4). In this method, the capillary is completely filled with sample (in high pH buffer), the cathode reservoir contains a high pH buffer (10 mM Tris/HCl, pH 8.5) and the anode reservoir a low pH buffer (500 mM Tris/HCl, pH

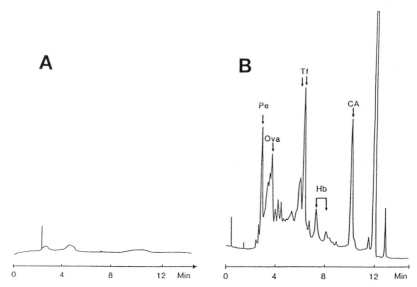

Figure 4 High-performance capillary zone electrophoresis of model proteins. (A) Prior to concentration; (B) following concentration toward a steep, nonbuffering pH gradient and a short mobilization by displacement electrophoresis. The large peak at 12 min may correspond to a moving boundary. (Reproduced from Ref. 112 with permission.)

2.5). Upon the application of an electric field, the sample components migrate toward the anode. Since the pH in the anode is 2.5, the proteins are unable to exit the capillary (they become protonated, and thus are repelled by the anode). Strong electrolytes (salt ions) are free to leave the capillary, and thus the sample is desalted as it is being concentrated. Once concentration has been achieved, a "mobilization" step (transient ITP) is performed by replacing the acidic buffer by a terminating electrolyte (30 mM glycine/NaOH, pH 10), reversing the polarity, applying voltage for a brief period of time. Finally the trailing electrolyte is replaced with a high pH buffer similar to the one contained inside the capillary and in the cathode reservoir. At this point, the separation continues as regular zone electrophoresis.

Preconcentration Using ITP with Hydraulic Counterflow

A combination of ITP and hydraulic forces was described by Hjertén et al. [107,112]. First, the capillary was filled with the sample dissolved in a leading electrolyte (15 mM HCl/Tris, pH 8.5), and a terminating electrolyte (100 mM glycine/NaOH, pH 8.5) was placed in the inlet reservoir. The outlet reservoir was filled with leading electrolyte. The outlet reservoir was elevated relative to the inlet reservoir to induce hydrodynamic flow counter to the direction of electrophoretic migration of the sample. When the electric field was applied, the sample was sandwiched between the leading and trailing buffers, and its migration was countered by the hydrodynamic flow, thus making the sample zones stationary (Figure 5). It is critical to match the counterflow velocity to the sample velocity to avoid loss of sample by migration into the electrode vessels. If the sample ions have mobilities intermediate to the trailing and leading electrolytes, they become concentrated at the boundary, and diffusion is eliminated (for reasons described previously for concentration by ITP). Upon completion of the concentration stage (about 15 min), the trailing buffer was replaced

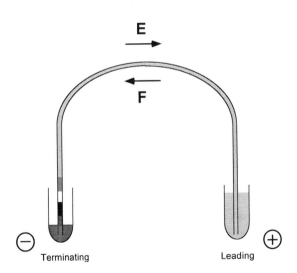

Figure 5 Schematic of transient ITP preconcentration with hydraulic counterflow. E indicates the direction of electrophoretic mobility during preconcentration; F indicates the hydraulic counterflow induced by level differences in the inlet (terminating) buffer reservoir and the outlet (leading) buffer reservoir.

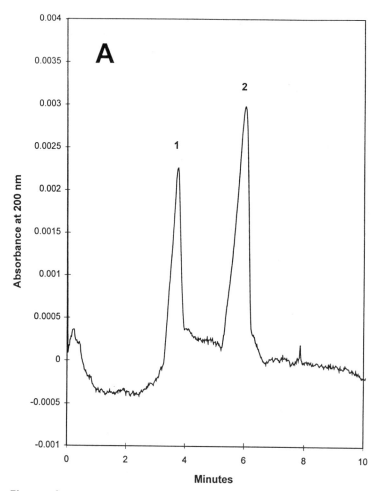

Figure 6   Comparison of the separation of β-lactoglobulin B (peak 1) and α-lactalbumin (peak 2) by (A) CZE and (B) CZE with ITP preconcentration and hydraulic counterflow. Leading buffer was 15 mM HCl + 100 mM Tris (pH 8.8) and terminating buffer was 100 mM alanine adjusted to pH 8.8 with Tris. Capillary was 24 cm × 75 μm (polyAAEE-coated), polarity was negative to positive, and detection was at 200 nm. CZE conditions were: injection, pressure for 2 psi*s (providing an injection zone length of 1 cm); leading buffer as electrolyte; voltage 10 kV. CZE with preconcentration conditions: capillary was prefilled with sample; preconcentration was carried out for 20 min at 1 kV with terminating buffer at the cathode and leading buffer at the anode; a height difference of about 2 cm between

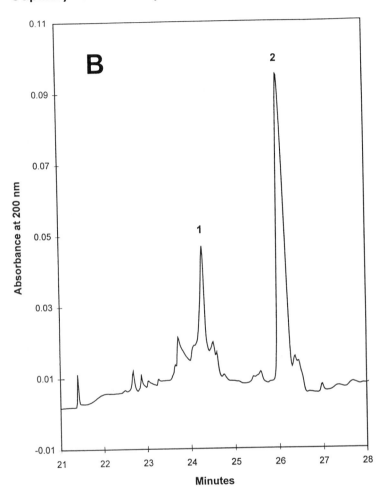

by leading electrolyte solution and the separation performed by CZE. This method can increase protein concentration over 50-fold when the proper conditions are selected (Figure 6). Precautions should be taken to avoid loss of the most rapid and slowest of the sample com-

---

terminating and leading buffer levels was used to generate gravity counter-flow. Following preconcentration, terminating buffer at the cathode was replaced by leading buffer at the same level as the anode reservoir and zone electrophoresis was carried out at 10 kV.

ponents during preconcentration. Quantitative differences for rapid and slow proteins were observed when the sample was diluted in the trailing instead of the leading buffer.

Preconcentration Through Adsorption to a
Chromatographic Support

Chromatographic supports, and specially affinity supports, can also be used to preconcentrate samples to be analyzed by CE [113]. This method has been demonstrated to work off-line as well as on-line. The main advantage of on-line preconcentration is the relative instrumental simplicity, whereas off-line preconcentrators offer more flexibility (and also the potential for higher concentration levels). Typical concentration factors are several hundred-fold. On-line solid phase concentration [114] has been used for the analysis of proteins such as metallothionein. The concentrator was constructed using a 5 mm length of polyethylene tubing (i.d. of 280 μm) fitted over the inlet end of a fused silica capillary. A 0.5 mm frit was made of fragments of a 0.45 μm nitrocellulose membrane packed into the concentration column using a small wire. Octadecyl packing material suspended in ethanol was then drawn into the concentrator by suction through the free end of the capillary. Another frit was placed at the end of the concentrator, and a piece of capillary (sample end) was fitted after the frit. After a brief conditioning step, the sample was injected into the column (up to 5 min) using low (0.5 psi) or high pressure (18 psi). The system was then flushed with buffer to remove any unbound protein, and to prepare the capillary for electrophoresis. The sample was then eluted with 33–50% acetonitrile using a small low-pressure pulse. Optimized conditions improved the sensitivity over 700-fold.

Chromatographic supports such as beads derivatized with C-18 are nonselective, and thus concentrate all proteins present in the sample. In instances when this is undesirable, selective preconcentration can be used. Cole and Kennedy [115] described the use of chromatographic supports derivatized with protein G and loaded with antibodies. The antibodies used had affinity for insulin and formed noncovalent interactions with protein G; thus the antibodies had to be replenished between analyses. The concentration column was a

30 cm × 150 μm capillary equipped with frit made out of 40–60 μm glass beads and immobilized by a short exposure to an open flame. To reduce protein adsorption, the capillary wall was treated with polybrene. On-line experiments used a flow-gated interface to couple the columns, whereas for off-line analysis the samples were first collected from the concentrator and then injected into the electrophoresis column. A 1000-fold concentration of insulin was achieved using this affinity chromatographic step.

Preconcentration to Regulate Sample Conductivity

The introduction of a zone of different conductivity into a capillary produces a heterogeneous distribution of the electric field. When the introduced zone is small, the electrical current usually "recuperates" to the level normally seen with the capillary filled with background electrolyte. But when the zone is of significant length (overloading), the migration time can be severely affected [116] (and in extreme cases the current drops to zero). Caution should be taken when using large bore and short columns, since they can be easily overloaded. For this reason it is necessary to preconcentrate diluted samples to increase their conductivity, thus improving reproducibility (and often resolution).

Hirokawa et al. [116] constructed a coupled device that allowed the concentration of diluted samples. The first section of the apparatus (ITP section) consisted of two electrode reservoirs filled with leading (10 mM HCl + β-alanine + 0.1% hydroxymethylcellulose, pH 3.5) and trailing (10 mM succinic acid + β-alanine + 0.1% hydroxymethylcellulose, pH 3.9) electrolytes, respectively. A separation column (fused silica capillary, 60 cm × 0.66 mm o.d. × 0.53 mm i.d.) was also filled with leading electrolyte, and was equipped with on-line, tandem UV and potential gradient detectors. An electric field was applied (constant current, 100 μA) until the sample components reached a second column (about 1 h), which was used as an interface to the CZE section of the system.

The sample was transported to a separation CZE column through the use of two valves that switched the electrical circuit to the CZE

Table 3 Buffer Systems Used for ITP Preconcentration of Proteins

Method	Leading	Trailing	Reference
ITP	120 mM HCl/Tris pH 8.9	190 mM taurine/ Tris pH 8.1	68
Transient ITP	20 mM triethylamine-acetic acid, pH 4.4	10 mM acetic acid	74
Coupled ITP	10 mM ammonium acetate-acetic acid, pH 4.8 + 1% Triton	20 mM ε-amino-caproic acid-acetic acid, pH 4.4	74
Coupled ITP	10 mM HCl/β-alanine, pH 3.5 + 0.1% MC	10 mM succinic acid/ β-alanine, pH 3.9 + 0.1% MC	80
Counterflow-ITP	15 mM HCl/Tris pH 8.5	100 mM glycine/ NaOH pH 8.5	70, 76
Transient ITP	10 mM ammoniun acetate, pH 3.6	50mM acetic acid, pH 3.1	75
Transient ITP	20 mM triethylamine/ acetic acid, pH 4.3	10 mM acetic acid	69
Transient ITP	Na-acetate (added to the sample)	20 mM β-alanine/ acetic acid, pH 4.3 (also used as BE)	69

section of the apparatus. The capillary and electrode reservoirs were filled with a homogeneous buffer.

To prove the usefulness of this system, the authors injected samples that differed widely in their conductivity and in the amount injected. Without the ITP preconcentration, the migration times varied nonlinearly with sample strength and amount injected. With the ITP-CZE apparatus, the migration times were remarkably constant.

A summary of buffer systems used for ITP preconcentration of proteins is provided in Table 3.

APPLICATIONS

The great number and variety of applications of CZE to the separation of proteins demonstrates the versatility of this CE mode. This section covers the major applications areas with a focus on recent developments in CZE separation of proteins.

Many reports describe the separation of proteins using MEKC conditions. The accepted mechanism for MEKC separation of small molecules is hydrophobic partitioning of the analyte into micellar structures, with SDS being the most widely used surfactant. The interaction of SDS with proteins is probably different. SDS micelles range from 3 to 6 nm in diameter, which is too small to accommodate molecules larger than about 5 kDa. However, proteins can bind tenaciously to SDS monomers and micelles via hydrophobic, hydrophilic, and electrostatic interactions. Protein-SDS binding is approximately stoichiometric with one SDS molecule bound per two amino acid residues; SDS-protein complexes therefore have approximately the same mass/charge ratio independent of protein size and are usually separated using sieving techniques (see Chapter 7). Evidence indicates that SDS-protein complexes consist of protein-enclosed micelles distributed along the protein chain [117]. Although protein behavior in SDS-containing buffers is qualitatively different than for small molecules, applications of MEKC-type conditions have been applied to many protein separations. In applications using uncoated capillaries, protein-wall interactions are eliminated because of the anionic character of SDS-protein complexes. In applications using coated capillaries with no electroosmotic flow, the high electrophoretic mobility of SDS-protein complexes can decrease analysis time. Use of MEKC-type conditions for analysis of proteins are discussed in the following sections, and has been recently reviewed by Strege and Lagu [118].

Enzyme Assays

Application of capillary electrophoresis to determination of enzyme activity typically involves off-line incubation of enzyme and substrate with timed injections of the reaction mixture into the capillary to separate and quantitate the low-molecular-weight substrate, intermediates, and product. An approach for in-tube enzyme assay using CZE has been developed by Regnier's group [119–122]. This system, termed electrophoretically mediated microassay (EMMA), is based on electrophoretic mixing of enzyme and substrate under conditions where mobilities of enzyme and product are different (Figure 7). The capillary was prefilled with all of the required components

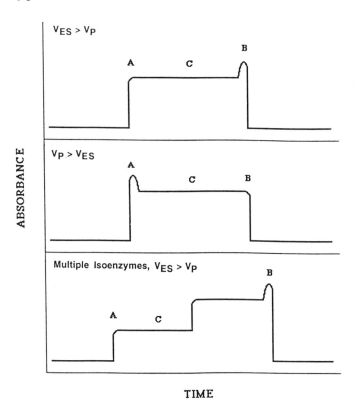

Figure 7 Predicted models showing various electropherograms in capillary electrophoretic enzyme assay. The moving velocities for enzyme-substrate complex (ES) and product (P) are (top) ES > P and (center) P > ES. A multiple isoenzyme form is shown in the lower panel with the moving velocity of their common product smaller than those of these isoenzymes. (Reproduced from Ref. 119 with permission.)

for the assay (buffer, substrate), and enzyme was introduced at the capillary inlet. Both product and enzyme were transported to the UV-vis detector, and product was detected at a selective wavelength where the enzyme did not interfere. The relative effective mobilities of enzyme and product were adjusted by manipulating the rate of EOF using covalent (e.g., epoxy polymer) or dynamic (e.g., nonionic surfactant adsorbed onto an octadecyl layer) coatings. The coatings also served to reduce adsorption of enzyme to the capillary wall. The analysis could be carried out in constant potential mode, in which enzyme was injected and transported down the capillary under constant voltage for the course of the analysis; sensitivity in constant potential mode could be increased by reducing the potential to increase incubation time. Highest senstivity was achieved by operation in zero potential mode; in this mode, enzyme was first injected and mixed with substrate, then voltage was turned off for a fixed incubation period to accumulate product. In the final step, potential was reapplied to transport product to the detection point. A limitation of the zero-potential mode was band spreading caused by diffusion of the product. This problem was solved by using a polyacrylamide gel-filled capillary cast under conditions such that the porosity of the gel reduced product diffision without introducing protein sieving effects. In this case incubation times of up to 2 h could be used in zero-potential mode. Electrophoretically mediated microassay was demonstrated for glucose-6-phosphate dehydrogenase (G6PDH, Figures 8 and 9) with glucose-6-phosphate + NADP/6-phosphogluconate + NADPH substrate/product (NADPH product detected at 340 nm), for alcohol dehydrogenase with ethanol + NAD/acetaldehyde + NADH substrate/product (NADH product detected at 340 nm), for β-galactosidase using o-nitrophenyl β-galactoside/o-nitrophenol substrate/product (product detected at 405 nm), and for alkaline phosphatase using p-nitrophenyl phosphate/p-nitrophenol substrate/product (product detected at 405 nm). Alkaline phosphatase (AP) was also assayed by EMMA using p-aminophenylphosphate as substrate and electrochemical detection with a prototype electrochemical detector installed in a modified commercial CE system [123]. Electrochemical detection provided an enzyme detection limit 10-fold lower in concentration than UV, but at the expense of greater complexity and limitations on separation conditions imposed by the detection system. Quantitative aspects of the G6PDH and AP reactions using UV detection of substrate and product was investigated by Zhao and Gomez [124]. Recently Regehr and Regnier [125] described enzyme assays using the EMMA approach coupled with chemiluminescence detection. In this system, enzymes yielding hydrogen peroxide as product could be assayed by detection of the accumulated product in an on-

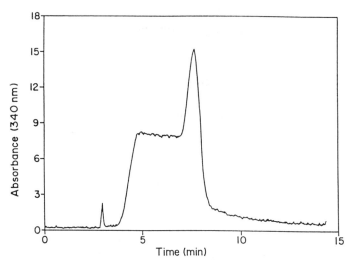

Figure 8 Electropherogram showing the accumulated product peak resulting from on-column assay of NADPH formed by the reaction of glucose-6-phosphate dehydrogenase with NADP. The running buffer contained all the reagents necessary to assay G-6-PDH; several minutes were allowed to elapse following injection of the G-6-PDH sample and the start of electrophoresis. (Reproduced from Ref. 119 with permission.)

line postcolumn reaction system employing luminol and horseradish peroxidase. Enzyme-amplified production of hydrogen peroxide provided detection limits of 7.7 amol and 120 zmol for galactose oxidase and glucose oxidase, while chemiluminescence detection of enzymatically consumed hydrogen peroxide yielded a detection limit of 15 zmol for catalase.

Saevels et al. [126] used the EMMA approach to determine the Michaelis constant for the adenosine deaminase conversion of adenosine to inosine using UV detection of the reaction product following electromigration of the substrate through a plug of enzyme; observed values agreed with previously reported values.

Fujima and Danielson [127] adapted an EMMA procedure for determination of lactate and pyruvate levels in serum using lactate dehydrogenase. The LDH-catalyzed conversion of pyruvate to lactate or the reverse reaction was monitored by production of NAD^+ (detected at 280 nm) or NADH (detected at 320 nm), respectively. The

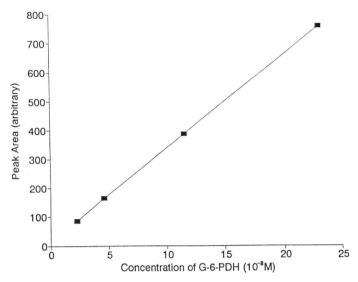

Figure 9 Relationship between the accumulated NADPH peak area and enzyme concentration in electrophoretically-mediated microassay of glucose-6-phosphate dehydrogenase activity. Reproduced from reference 119 with permission.

latter reaction required periodic washing of the capillary with enzyme + hydrazine to remove accumulated substrate. The 1 min assay provided high sample throughput, and enzyme lifetime was extended by stabilization with PEG. A limitation of the method was the requirement for an ultrafiltration step to remove interfering proteins from the serum sample.

Recently Sun and Hartwick [128] described a different approach for in-tube monitoring of enzyme activity. A multipoint detection scheme was used which employed a capillary with four windows positioned along the capillary length and mounted in a common detector cell. The electrophoresis buffer contained the enzyme, and conversion of substrate to product could be monitored at four time points during migration; an internal standard was included to compensate for varying internal capillary diameters at the window positions. This system was used to monitor the adenosine deaminase conversion of adenosine to inosine.

Analysis of lactate dehydrogenase (LDH) activity in single eryth-rocytes and lymphocytes has been studied by Yeung's group [129,130]. Individual cells were introduced into a capillary filled with phosphate buffer at physiological pH containing lactate and NAD^+. Following cell lysis, isoenzymes were separated electrophoretically and the activity of each species was determined by laser-induced fluo-rescence detection of the NADH product. For red blood cells, lysis occured spontanteously by introduction of the cell into the hyper-tonic electrophoresis buffer [129]. Lysis of lymphocytes was accom-plished with an induced electromagnetic field using a tesla coil posi-tioned at the capillary inlet [130].

Affinity Capillary Electrophoresis

Affinity capillary electrophoresis (ACE) is a method for studying receptor-ligand binding in free solution using capillary electrophore-sis. The technique depends upon a shift in the electrophoretic mobil-ity of the receptor upon complexation with a charged ligand. If asso-ciation and dissociation have slow kinetics, the sample will be resolved into bound and free components and binding constants can be calculated from the peak areas at different ligand concentrations in the electrophoresis buffer. In the case of fast kinetics, only one peak will be observed and affinity is determined by the change in migration time due to variations in time spent in bound and free states during migration through the capillary. The mobility of the affinity complex will shift from that of free receptor to a maximum at full saturation, and the relationship between mobility shift and ligand concentration can be used to determine the receptor–ligand associa-tion constant. Affinity capillary electrophoresis has been reviewed by Shimura and Kasai [131].

Affinity capillary electrophoresis has been applied to the study of protein–drug, protein–protein, protein–carbohydrate, and protein–nucleic acid interactions. Conventional methods of studying receptor–ligand interactions include equilibrium dialysis using radiolabeled species, UV or fluorescence spectroscopy, nuclear magnetic resonance, and differential scanning calorimetry. Compared to these methods, ACE offers several advantages including requirement for very small amounts of analytes, speed, automation, and the ability to determine ligand interactions with multiple receptors. In addition, pure receptor preparations or accurate concentration values are not required since only migration times are measured. In a typical ACE experiment, the receptor is injected into a capillary containing free ligand at a variety of concentrations; depending upon the kinetics of the on and off processes, incremental shifts in migration times are observed, and Scatchard analysis of migration time shifts (usually normalized to a neutral EOF marker or a nonbinding reference protein) in response to ligand concentration is used to estimate the ligand–receptor binding or dissociation constant. A model system for measuring protein–ligand interactions using carbonic anhydrase as the receptor and arylsulfonamides as ligands (Figure 10) has been well character-

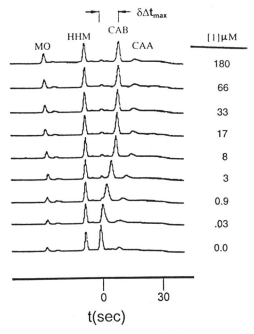

Figure 10 Affinity capillary electrophoresis of bovine carbonic anhydrase B in Tris-glycine buffer (pH 8.4) containing various concentrations of a charged aryl sulfonamide as affinity ligand. Horse heart myoglobin (HHM) and mesityl oxide (MO) were used as internal standards. (Reproduced from Ref. 132 with permission.)

ized [132–134] and the values for binding constants determined by ACE were in good agreement with those determined by other techniques. In cases where the receptor–ligand dissociation time is of the same magnitude as its migration time, analysis of the extent of peak broadening can be used to estimate K_{on} and K_{off} values. Binding constants for neutral ligands can be estimated from shifts in the migration time of charged ligand–protein complexes in the presence of increasing amounts of a competing neutral ligand. Several limitations must be addressed when developing an ACE system: wall interactions can interfere with the separation process and affect the receptor–ligand complex equilibrium, while use of additives to reduce wall effects may also alter receptor–ligand interaction.

Estimation of binding constants by changes in protein mobility due to ligand complexation can be problematic if the magnitude of EOF is affected by varying ligand concentration. The ligand may affect EOF by interacting with the wall (especially cationic ligands) or by changing the viscosity or thermal conductivity of the electrophoresis buffer. Gomez et al. [135] demonstrated that mobility corrections using a neutral marker (mesityl oxide) could compensate for variable EOF to yield accurate estimates of binding constants.

High-performance size exclusion chromatographic techniques for determining protein-drug binding have been adapted for capillary zone electrophoresis by Kraak et al. [136]. Using warfarin binding to bovine serum albumin as a model system, the Hummel–Dreyer method, vacancy peak method, and frontal analysis were compared. Frontal analysis was found to be the most reproducible and yielded binding constant estimates in closest agreement with previously reported values. The frontal analysis method was compared to ELIZA determination of hapten–antibody association constants [137] and was found to provide an easily automated method of high precision. However, the antibody volume and time required were much higher than ELIZA techniques, and matrix interferences were a problem. Poor sensitivity using UV detection limited acquisition over the full range of free hapten concentration, a particular problem for high association constants.

Affinity complexation can be detected indirectly, as demonstrated by Busch et al. [138]. In this technique, termed vacancy affinity capillary electrophoresis (VACE), the capillary is filled with buffer containing a mixture of receptor and ligand, a plug of buffer is injected at the capillary inlet, and voltage is applied. Differential migration of ligand and receptor through the buffer zone produces zones deficient in ligand and receptor, resulting in the appearance of two negative peaks at steady state. In the VACE experiment, the concentration of ligand or receptor is varied and the mobilities of the two peaks relative to a neutral marker is monitored. The magnitude of the peaks provides information about the degree of complexation, e.g., the receptor–deficient peak reflects the concentration of free receptor and the ligand–deficient peak indicates the level of free ligand. Similarly, the shift in mobility of each peak provides information on the receptor–ligand association constant and the number of receptor binding sites. Using vancomycin and the dipeptide N-acetyl-D-alanyl-D-alanine as the receptor-ligand pair, these authors compared the ACE and VACE approaches and found satisfactory agreement of the measured association constants.

Affinity capillary electrophoresis has been used to characterize protein–sugar interactions by Honda et al. [139]. Lactobionic acid was used as the charged ligand to determine association constants for three β-galactoside–specific lectins. These studies were performed near the isoelectric points of the lectins which, in the absence of ligand, migrated with a neutral EOF marker. In the presence of the acidic disaccharide ligand, migration of the sugar–protein complex was retarded and Scatchard analysis of the data yielded association constants essentially identical to those obtained from equilibrium dialysis. Kuhn et al. [140] used the same approach to study the interaction of fucose 1-phosphate with three lectins isolated from the asparagus pea, and to determine the effect of calcium ion on lectin–carbohydrate complexation.

Affinity capillary electrophoresis has been used to compare the heparin-binding of lactoferrin obtained from human milk (M-LF) and human granulocytes (G-LF). These two lactoferrins are similar in terms of primary sequence, glycosylation, and iron-binding capacity. Heegaard and Brimnes [141] used ACE in a coated capillary with phosphate buffer (pH 8.2) supplemented with varying concentrations of heparin to characterize the binding properties of the two lactoferrins. Lactoferrin mobility was observed to increase with increasing heparin concentration. The similarity in the degree of peak mobility shift and changes in peak shape suggested similar heparin binding behavoir for the two proteins.

Heegaard et al. [142] also studied DNA binding to an anti-DNA monoclonal antibody by migration-shift affinity capillary electrophoresis. The minimal binding size was >16 bases, and the interaction of double-stranded DNA was 10-fold stronger than for single-stranded DNA. Binding was pH-dependent, and strongly decreased as a function of ionic strength.

Application of ACE to determination of protein-ligand binding stoichiometry for charged ligands was described by Chu et al. (143). Determinations for weak-binding systems (fast off rate) were performed by including the ligand in the electrophoresis buffer as well as the sample at concentrations well above K_d. At a fixed receptor concentration, the point at which the free ligand response in the electropherogram passed through zero was used to estimate the concentration of bound ligand. Aryl sulfonamide binding by carbonic anhydrase ($n = 1$) was used as a model weak binding system. Determinations for strong-binding (slow off rate) systems were performed simply by varying the ligand concentration at fixed receptor concentration and determining the total ligand concentration at which free ligand was detectable in the electropherogram; partially bound intermediates could be detected as peaks migrating between unbound receptor and fully bound ligand–receptor complexes. Examples of strong-binding systems were human serum albumin/anti-HSA ($n = 2$) and streptavidin/biotin ($n = 4$). In the latter case, partially bound intermediates could not be resolved due to insufficient charge differences. By subsituting a biotinylated 15mer oligonucleotide for biotin, enhanced charge differences permitted resolution of all intermediates.

The possibility of using CZE to monitor antibody–antigen complexation was described several years ago [144]. Recently Reif et al. [145] reported the use of fluorescein isothiocyanate-labeled protein G as an affinity ligand in conjunction with laser-induced fluorescence detection for quantitation of FITC-protein G-antibody complexes. The technique was applied to determination of IgG in serum using FITC-protein G or FITC-protein G tagged with anti-IgG. Very rapid separation conditions were used to minimize dissociation of antibody–antigen complexes. One drawback of this approach was the broad, multicomponent FITC-protein G peak resulting from antigen tagged with variable numbers of fluorophores, which interfered with quantitation of the antibody–antigen complex at low IgG concentrations.

Pédron et al. [146] used ACE to monitor the coupling of the plant hormone abscisic acid (ABA) to bovine serum albumin. Mobility shifts of the ABA–BSA complex were used to follow the kinetics of coupling, with the magnitude of the shift correlated to the molar coupling ratio and the peak shape of the complex providing information on the distribution of hapten–protein conjugates according to their coupling ratios.

An ACE assay for insulin has been developed by Kennedy's group [147] which is based on mixing FITC-labeled insulin with monoclonal anti-insulin in the presence of unlabeled insulin. Determination of insulin in a sample was perfomed by separation of bound and free species and measurement of the ratio of free FITC-insulin to an internal standard; the technique has been applied to measurement of antibody–antigen dissociation constants [148]. The competitive insulin immunoassay has been automated [149] to determine the insulin concentration in a flowing stream by combining sample, FITC-insulin, and antibody streams in a mixing cross, and flow-gating segments of the reaction mixture into a separation capillary. Separation of free from bound FITC insulin was accomplished within 3 s, and the system was applied to determination of the insulin concentration in individual islets of Langerhans.

Recent progress in ultraminiaturization of capillary electrophoresis separations on microfabricated devices offers significant opportunities for chip-based immunoassays in clinical diagnostics. However, one of the challenges is minimizing protein-wall interactions. Since electroosmotic flow is used as the transport mechanism in chip-based devices, coatings which suppress protein adsorption and EOF are impractical. In adapting commercial immunoassay systems for theophyline to a chip-based system, Chiem and Harrison [150] used a Tricine buffer with sodium chloride and Tween 20 as additives to suppress antibody adsorption without eliminating EOF.

ACE has been used to assess the binding of the immunosuppressant deoxyspergualin (DSG) and DSG analogs to the heat shock proteins Hsc70 and Hsp90, which function as molecular chaperones. Nadeau et al. [151] employed uncoated capillaries using a Tris-glycine buffer at pH 8.3 to determine K_d values for DSG binding to murine, bovine, and trypanosomal Hsc70 proteins. Liu et al. [152] compared DSG binding to human T cell Hsc70 using uncoated and coated capillaries, and demonstrated that K_d values obtained with coated capillaries at pH 2.8 were 8- to 9-fold lower than those obtained with uncoated capillaries at pH 5.3 or 6.95.

Separation of human and bovine serum albumin was shown to be improved by addition of antisteroidal inflammatory drugs as affinity ligands to the electrophoresis buffer, and this system was used to estimate the binding constants of HSA and BSA for ibuprofen and flurbiprofen [153]. This study employed coated capillaries, so no correction for EOF was required; α-lactoglobulin (which did not bind the ligands) was used as an internal reference standard.

Chadwick et al. [154] studied the irreversible binding of N-oxidized metabolites of the antiarrhythmic drug procainamide to hemoglobin using CZE with uncoated capillaries and a 0.1 M sodium tetraborate buffer at pH 8.5. Estimates of the number of protein binding sites and the protein–ligand association constant were obtained by determinations of amount of free ligand as a function of total ligand concentration, and the values obtained were in close agreement with those determined using flow injection analysis with electrochemical detection of free ligand.

CZE has been used to monitor competitive adsorption of human serum albumin onto polyisobutylcyanoacrylate nanoparticles coated with human orosomucoid [155]. In this study, CZE was used to detect the amount of HSA and orosomucoid in the supernatant solution following incubation of particles with the competitive ligand, and to quantitate both proteins following desorption from the particles with detergent. Improved resolution of HSA and orosomucoid was obtained by the addition of 25 mM SDS to the electrophoresis buffer; the detergent bound strongly to HSA (increasing its effective mobility) but bound inefficiently to the highly glycosylated orosomucoid. Complexation with SDS also improved HSA peak shape, presumably by reducing electrostatic interaction with the negatively charged capillary wall.

Chen and Sternberg [156] developed an immunoassay for digoxin using CE with laser-induced detection. Digoxigenin was conjugated with a 10mer oligonucleotide tagged with tetramethylrhodamine (TMR). The oligonucleotide served to modulate the charge of the antigen, and the TMR provided high-sensitivity fluorescence detection using an argon ion laser. Free and anti-Fab-bound conjugate were separated by CZE, and the relative peak heights of the two species in the presence of competing unlabeled digoxin were used to constuct a calibration curve for determination of digoxin levels.

Okun and Bilitewski [157] have also used the technique of ligand modification by conjugation with an oligonucleotide to minimize adsorption and move the complex away from interfering sample components. In this study, a biotinylated 14mer oligonucleotide was used to assay the level of the biotin–binding protein actinavidin. The surfactant SDS was added to the electrophoresis buffer at a concentration slightly above the CMC to prevent protein adsorption to the capillary wall; this reduced the actinavidin binding capacity by about 12% but did not interfere with quantitation of the protein in culture liquids of *Streptomyces chromofuscus*.

Use of CZE to separate bound from free antigen in a competitive immunoassay for angiotensin II was described by Pritchett et al. [158]. The antigen was labeled with cyanine dye and detected using by laser-induced fluorescence using a helium–neon laser. Separations were performed in an uncoated capillary using a borate buffer at pH 10.25. Angiotensin was analyzed in serum samples with a concentration limit of detection of about $10^{-10} M$.

Affinity capillary electrophoresis has been used to detect scrapie prion protein in sheep brain preparations [159]. In this study, fluorescein-labeled synthetic prion epitope peptides were resolved from peptide–antiprion antibody by CZE with LIF detection. Using a 200 mM Tricine buffer (pH 8.0) and a 20 μm i.d. uncoated capillary, the system formed the basis for a competitive assay of brain preparations from normal and scrapie-infected sheep.

Frokiaer [160] used ACE to study the binding of plant protease inhibitors. The binding of trypsin or chymotrypsin inhibitors was determined by formation of inhibitor–protease complexes which could be resolved from free enzyme and inhibitor using a taurine buffer with cholate additive. Binding of inhibitors to monoclonal antibodies was studied using the same system.

A novel ACE system based on immobilized metal ion affinity has been described by Haupt et al. [161]. In this system a polymer (monomethoxy polyethylene glycol, m-PEG) carrying copper-bound immobilized iminodiacetic acid (IDA-Cu[II]) functions was added to the run buffer. Migration of proteins in the buffer–polymer solution was modulated by interaction of available histidine residues with the IDA-Cu(II) ligand. To achieve specific histidine–ligand interactions, the buffer pH was maintained between 6.5 and 7.5, and to reduce nonspecific interactions a relatively high buffer ionic strength was needed. This in turn required use of 20 μm i.d. capillaries to reduce Joule heat. By measuring the retardation of migration time as a function of ligand concentration, the dissociation constants of ribonucleases A and B were determined, and the inhibition constants could be calculated by varying the concentration of a competing ligand (imidazole) in the run buffer. Application of the immobilized metal affinity capillary electrophoresis (IMACE) technique to an acidic protein (kallikrein) required use of a hydrophilic neutral coated capillary with reversed polarity.

Chiral Separations

Proteins have been widely used as stationary phases for separation of chiral compounds in HPLC and their application as chiral selectors in capillary electrophoresis has been extensively investigated (see Refs. 162–165 for recent reviews). Use of capillary electrophoresis with protein chiral selectors offers many advantages including simplicity (no requirement to attach the protein to a support), flexibility (any soluble protein can potentially be used, and a given protein may interact with a variety of analytes), and capacity (the protein is in free solution). Since separation is achieved by interaction of the analyte with a protein pseudophase, this technique can therefore be considered a mode of electrokinetic chromatography. To achieve adequate resolution, the mobility of the protein-bound complex must be significantly different from that of free analyte. Resolution of an enantiomeric pair arises from differential migration of enantiomer–protein complexes (usually in the presence of EOF) and since kinetics of complex formation and protein mobility can be highly sensitive to pH, adjustment and control of the buffer pH is key to optimization of a method. Proteins used as chiral selectors in CZE are described in Table 4.

Use of proteins as pseudophases for chiral separations has two major limitations. As in applications of CE for separation of protein analytes, use of proteins as chiral selectors is limited by protein-wall interactions, which can effect EOF and reduce analytical reproducibility. Proteins also exhibit strong absorbance in the low UV, and detection sensitivity is extremely limited due to high background absorbance. Detection can be performed at longer wavelengths, but this limits application of the technique to compounds with absorbance maxima in the high UV range. For these reasons, several investigators have employed columns packed with proteins coupled to gels or particles in which the detection segment is free of packing. However, the magnitude of EOF flow in packed columns is low

Table 4 Characteristics of Proteins Used as Chiral Selectors

Protein	M_r	pI	% Carbohydrate
α_1-acid glycoproten	44,000	2.9–3.2	45
Avidin	70,000	10–10.5	20.5
Bovine serum albumin	67,000	4.7–4.9	—
Cellobiohydrolase I	60,000–70,000	3.9	6
Conalbumin	77,000	6.1–6.6	—
Fungal cellulase	60,000–70,000	3.9	6
Human serum albumin	68,000	4.7	—
Ovomucoid	28,000	3.8–4.5	20

compared to open tubular columns, and packed columns must be operated at low ionic strengths or low field strengths to avoid problems in bubble formation.

Birbaum and Nilsson [166] immobilized bovine serum albumin by cross-linking the protein with glutaraldehyde in situ within the capillary to form a gel; the gel terminated before the detection point which enabled detection of analytes at 214 nm; separation of D- and L-tryptophan was demonstrated.

Barker et al. [167] employed bovine serum albumin (BSA) as an enantioselective buffer additive for separation of stereoisomers of leucovorin, a folate derivative used in cancer therapy. Poor reproducibility and capillary lifetime were observed with uncoated capillaries, presumably due to protein-wall interactions. Use of a polyethylene glycol-coated capillary resulted in almost a 10-fold improvement in reproducibility and capillary lifetime. In cases where BSA and the analytes had similar mobilities or where the analytes exhibited weak binding to the protein, resolution of enantiomers was poor. These authors demonstrated [168] that the addition of a UV-transparent polymer additive (e.g., 5% 2,000,000 M_r dextran) selectively retarded migration of the BSA additive and allowed separation of enantiomers of several drugs and amino acid derivatives which could not be resolved with BSA in the absence of the polymer. In a later refinement of this approach, these authors [169] immobilized BSA on the dextran polymer using cyanogen bromide. The mobility of the BSA–dextran polymer was very low relative to that of the analytes, permitting rapid separations of enantiomers in short capillaries. An additional advantage of this approach was the ability to vary the selector phase ratio simply by dilution with underivatized dextran. Vespalec et al. [170] investigated human serum albumin as a buffer additive for chiral separations of amino- and carboxylic acids. They observed that slow changes in the enantioselectivity of the buffer system could be avoided by heating of the albumin solution prior to use. At high albumin concentrations, adsorption of protein to the wall of fused silica capillaries caused strong decreases in EOF and consequently long analysis times; this was remedied by using linear polyacrylamide-coated capillaries. Evaluation of serum albumins from seven different animal sources for the separation of enantiomers of ofloxacin (an antibacterial quinoline) demonstrated that enantio-selectivity was variable for the different proteins [171,172]. Interestingly, successful resolution was achieved with BSA although protein–drug binding was observed with human serum albumin (HSA). Chemical modification with palmitic, glucosamide, or acetyl groups reduced or eliminated BSA enantioselectivity; addition of other drugs to the electrophoresis buffer provided information about the ofloxacin binding sites on native BSA. Lloyd et al. [165] compared the use of HSA as a chiral selector immobilized on packed beds or in free solution for electrochromatographic separations of a variety of enantiomers. Capillaries with HSA phases on 7 μm dp silica exhibited low EOF and lower efficiency compared to open tubular systems. Free solution studies employing dextran as an additive to increase the mobility differences of free vs. bound ligand suggested that the polymer additive affected protein–ligand binding.

A widely used protein for chiral separations in HPLC is α_1-acid glycoprotein (ACP, orosomucoid), and its use in CE separations has been investigated [173]. AGP is a strongly acidic glycoprotein (pI 2.7) containing 47% carbohydrate; both the amino acids of the primary sequence and the sugars of the carbohydrate moeties can participate in chiral interactions. Capillaries packed with AGP bonded to 5 μm silica particles were used for chiral separation of a group of β-blockers, barbituates, and nonsteroidal anti-inflammatory drugs. The effects of buffer pH and ionic strength, organic modifier type and concentration, and field strength on EOF and enantioselectivity were studied. Separation efficiencies were generally higher than those obtained with HPLC columns packed with AGP phases, but lower than those observed in CE separations using protein additives in free solution.

Ovomucoid is an acidic glycoprotein obtained from egg white which has been used as a chiral stationary phase in HPLC. Ishihama et al. [174] investigated the use of ovomucoid in free solution for electrokinetic chromatographic separation of several drugs. Poor results were obtained using uncoated capillaries due to protein-wall interactions, but the use of PEG-coated capillaries, hydroxypropyl-cellulose as a buffer additive, or 2-propanol as an organic modifier improved reproducibility and efficiency.

Avidin has been used as a chiral selector for a variety of acidic compounds [175]. Avidin is a very basic protein (pI ~ 10), and required use of coated capillaries to reduce protein-wall interactions. In the absence of electroosmotic flow, avidin migrated in a direction opposite to that of the analytes, and optimization of the buffer pH was required to achieve enanatiomeric resolution with reasonable analysis times. As is often observed when using proteins as chiral selectors, the slower-migrating member of an enantiomeric pair exhibited significant peak asymetry (perhaps due to slow binding kinetics); increasing the capillary temperature improved peak shape. In some cases where very strong analyte–protein interaction reduced analyte mobility, interaction could be reduced by addition of an organic modifier (e.g., 2-propanol) to the electrophoresis buffer.

The enantioselectivities of four proteins (BSA, ovomucoid, orosomucoid, and fungal cellulase) used as buffer additives were compared by Busch et al. [176]; five pairs of enantiomers (tryptophan, benzoin, wafarin, pindolol, promethazine, and disopyramide) were used as analytes. All four proteins except ovomucoid were able to resolve enantiomers of some but not all of the analytes, and selectivity was affected by buffer pH and the type and concentration of organic modifer used (1-propanol, dimethyloctylamine). Wistuba et al. [177] compared the effectiveness of BSA, AGP, ovomucoid, and a mixture of α-, β-, and γ-casein as chiral selectors for six 2,4-dinitrophenyl amino acid derivatives. BSA provided best overall resolution of the six analytes, while the casein preparation provided best resolution of D- and L-glutamic acid.

Use of a soluble protein as a chiral selector under conditions which permitted low UV detection was described by Valtcheva et al. [178]. The capillary was partially filled with buffer containing cellobiohydrolase, leaving the detection window of the capillary protein-free; separations were carried out at a buffer pH such that the injected analytes (a series of β-blockers) migrated toward the detector (cathodic end) while the enzyme migrated toward the capillary inlet (anodic end). Electroosmotic flow was eliminated by using a linear polyacrylamide-coated capillary, and hydrodynamic flow was eliminated by introducing an agarose plug at the capillary end. Enantiomeric resolution required high buffer ionic strength (0.4 M phosphate), necessitating the use of low field strengths to prevent excessive heating. In addition, the use of high concentrations of an organic modifier (up to 30% isopropanol) was necessary to prevent peak broadening due to hydrophobic interactions of the analytes with the enzyme. Involvement of the enzyme active site in chiral recognition was evidenced by impairment of enantioselectivity by the enzyme inhibitor cellobiose, and inhibition of enzyme activity by one of the analytes (propanolol). Tanaka and Terabe [179], using a similar approach, developed a method for chiral separations which did not require use of an agarose plug in the capillary and which could be fully automated using a commercial CE system (Figure 11). In this method,

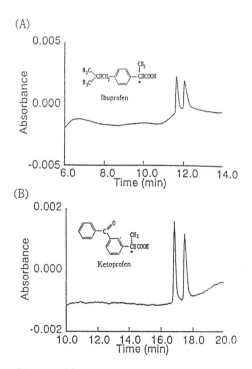

Figure 11 Enantiomeric separation of racemic (A) ibuprofen and (B) ketoprofen using partial filling of the capillary with 25 μM avidin in 50 mM phosphate buffer (pH 6.0) containing 10% IPA. Separation performed at −12 kV in a 31.5 cm (effective length) × 50 μm coated capillary (Bio-Rad). (Reprinted from Ref. 164 with permission.)

the capillary was prefilled with the electrophoresis buffer, then buffer containing the chiral selector was introduced at low pressure to form a separation zone occupying only the segment of capillary before the detection point. The separation conditions (pH and polarity) were adjusted so that the analytes injected at the capillary inlet migrated through the segment of chiral selector, while the protein had low mobility and did not migrate significantly during the analysis. This approach offered several advantages: high protein concentrations could be used to effect chiral separations without compromising sensitivity, only small volumes of protein are needed to fill the separation segment of the capillary, and automation of the method provided excellent reproducibility. Bovine serum albumin, α_1-acid glycoprotein, ovomucoid, and conalbumin were use as chiral selectors in this study. More recently, these authors have investigated the use of α_1-acid glycoprotein (AGP) as a chiral selector with partial capillary filling in detail [180]. The pH of the buffer was raised to enhance electrostatic interaction of basic analytes with the acidic protein selector; this also increased the mobility of the protein selector toward the anode. By optimizing the pH of the buffer in the pH 5–6 range, varying the concentration of AGP from 50 to 1000 μM, and (in some cases) adding organic modifiers, 29 basic drugs could be resolved into their enantiomers. Preparations of AGP from three different commercial sources were compared; although there was usually no difference in selectivity, some drug enantiomer pairs could not be resolved using AGP from a particular source. The partial-filling technique could accurately detect as little as 0.3% of a contaminating isomer.

Analysis of Protein Folding

Capillary electrophoresis has been shown to be a valuable tool for analysis of protein folding. Unlike techniques such as chromatography and gel electrophoresis, capillary zone electrophoresis is performed in free solution and migration is a function of the intrinsic properties of the molecule. The ability of CZE to distinguish different folding states of a protein depends upon changes in solvent-accessible charge, and the migration rate thus reflects a cross section of the conformational states. Moreover, peak shape can provide information on the distribution of the protein among folding states. Capillary electrophoresis separations are performed in a short time frame, permitting detection of short-lived unfolding intermediates which might not be observed using chromatography or gel electophoresis. Rush et al. [181] first described the effect of thermally induced conformational changes on migration behavior of α-lactalbumin, observing a sigmoidal dependency of the viscosity-corrected mobility on temperature, with a transition temperature which agreed closely with that determined by intrinsic fluorescence measurements. Strege and Lagu [182] used CZE to monitor reformation and interchange of disulfide bonds during reoxidation of reduced trypsinogen. In this study, capillary electrophoresis was performed under low pH conditions to minimize protein-wall interactions for this basic protein. A population of refolding intermediates distributed between native and unfolded trypsinogen was resolved, and resolution was improved by addition of ethylene glycol and sieving polymers. These authors also monitored the transition of bovine serum albumin from the native folded state to the unfolded state using CZE in the presence of increasing amounts of urea [183], and the resulting plot of EOF-corrected migration vs. urea concentration was similar to urea denaturation profiles obtained with other techniques. Using capillary electrophoresis in the presence of 0–8 M urea, Kilár and Hjertén [184] detected intermediate unfolding states of transferrin as distinct peaks, and were able to resolve unfolding intermediates for each of the five transferrin glycoforms (2-, 3-, 4-, 5-, and 6-sialotransferrin). Hilser et al. [185] monitored the migration behavior of lysozyme as a function of capillary temperature and observed a sigmoidal behavior characteristic of a protein unfolding transition. Calculation by van't Hoff analysis of the transition temperature, entropy, and enthalpy of unfolding yielded values in close agreement with those determined by differential scanning calorimetry, and confirmed that the temperature-dependent decrease in electrophoretic mobility represented a two-state thermal denaturation. As observed previously for α-lactalbumin [181], lysozyme exhibited sharp peaks at low and high temperature, but migrated as a broad peak at the intermediate temperatures with maximum broadening at the transition midpoint (Figures 12 and 13). This behavior was consistent with

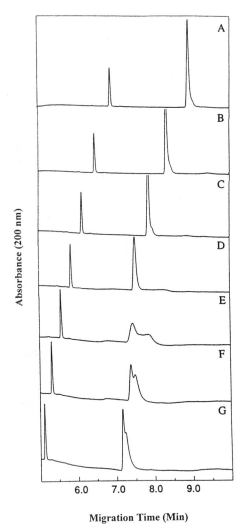

Figure 12 Capillary electrophoresis of lysozyme at (A) 37°C, (B) 42°C, (C) 47°C, (D) 52°C, (E) 57°C, (F) 62°C, and (G) 67°C. The first peak corresponds to a 7-residue peptide mobility standard and the second peak is lysozyme. While a plot of mobility vs. temperature was linear over the temperature range for the peptide standard, the plot for lysozyme was sigmoidal in the transition region of 50–60°C, and peak broadening was observed. The more slowly migrating peak was attributed to interactions of the protein with the fused silica capillary. (Reproduced from Ref. 185 with permission.)

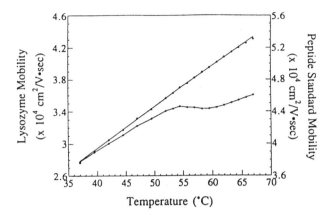

Figure 13 Electrophoretic mobility values obtained at various temperatures for a peptide standard RKRSRKE (closed triangles) and lysozyme (closed circles). (Reproduced from Ref. 185 with permission.)

fast interconversion of folded and unfolded states during transition, with mobility of the peak determined by the equilibrium population of molecules in each state. The appearance of a second minor peak in the transition region was interpretated as a slow-refolding subpopulation of molecules, perhaps arising from protein interaction with the uncoated capillary surface. This interpretation was supported by the disappearance of the minor peak when the experiments were repeated using polyacrylamide-coated capillaries. In a recent review, Hilser and Freire [186] simulated protein migration behavior in response to thermally induced unfolding to demonstrate the utility of CZE for quantitative analysis of protein folding. Two cases were treated, a slow-time regime in which interconversion is much slower than electrophoresis time (yielding distinct peaks for folded and unfolded species) and a fast-time regime in which interconversion is much faster then electrophoresis time and the peak represents a time-averaged combination of component species. In the fast-time regime, apparent thermodynamic parameters can be calculated for two-state unfolding and kinetic rate constants obtained from analysis of peak width. In the slow-time regime, true thermodynamic parameters can be obtained and characterization of unfolding transitions is not limited to the two-state case.

Capillary electrophoresis has been used as a confirmatory technique to assess the confirmational states of proteins eluted from reversed-phase HPLC columns. Bishop et al. [187] demonstrated that Rnase A injected in four conformational states (native, partially denatured, completely unfolded, and completely unfolded and disulfide-reduced) exhibited a primary peak which eluted later in the denatured and denatured + reduced sample. Particle beam FT-IR spectroscopy revealed increased amounts of β-sheet structure in the primary peak fractions from the denatured and denatured + reduced samples. They interpreted these results as loss of structure during the chromatographic process, and refolding during desorption and elution. Rnase A prepared with increasing concentrations of urea were analyzed by CZE and migration time was observed to increase during the transition from the native state (<5 M urea) to fully the unfolded state (8 M urea), presumably due to increase frictional drag of the denatured protein. When eluted fractions from RPLC were subjected to CZE analysis, RPLC peaks which exhibited β-sheet structure migrated between native and fully denatured RNase, confirming the existence of secondary structure in these partially refolded forms.

Kats et al. [188,189] used MEKC conditions to resolve pH-dependent isoforms of a monoclonal chimeric antibody. After incubation of the antibody at pH values from 2 to 12, it could be resolved into five species representing different structural states. The quantitative distribution among the five species depended upon the pH: low pH produced only a slow-migrating species, basic pH produced a fast-migrating species, while intermediate pH values populated all five states in varying amounts. Surprisingly, CD spectroscopy revealed little difference in secondary structure under conditions (neutral pH) where several conformers were resolved, while conditions shown to significantly change secondary structure (low pH, heat treatment with chaotropic agents) did not produce species resolvable by capillary electrophoresis. Apparently the more subtle changes at intermediate pH altered protein–surfactant interactions allowing resolution of the conformers. In a later study, the same authors used MEKC conditions to study structural changes in a fusion protein consisting of variable regions of the same antibody linked to the *Pseudomonas* exotoxin PE40 (190). The fusion protein was treated with varying concentrations of guanidine hydrochloride (which destroys all ordered structure) or trifluorethanol (which stabilizes secondary structure such as α-helices but destabilizes tertiary and quaternary structure). Separation of fusion protein conformers required replacement of SDS with cholic acid in the MEKC buffer. In the presence of guanidium hydrochloride, the characteristic three-peak profile of the fusion protein conformers collapsed into a single fast-migrating broad peak, and the CD spectra indicated loss of ordered structure as the protein was converted to random coil with increasing denaturant concentration. In contrast, trifluoroethanol produced shifts in the isoform populations to more slowly migrating species, while CD spectra showed increasing α-helicity and decreasing tertiary structure.

Refolding of the tailspike protein of phage P22 was monitored with capillary electrophoresis by Fan et al. [191]. This is a 666-amino acid protein which exists as a trimer in its native state and assembles onto the phage capsid to form the cell recognition and attachment apparatus of the phage. The native trimer is thermostable and resistant to SDS, but in the presence of high concentrations of urea ($7\ M$) and acidic conditions is denatured to random-coil monomers. Refolding occurs upon dilution into neutral pH buffers with low urea. This process occurs slowly at low temperatures ($10°C$) and was monitored using CZE conditions (uncoated capillary, 25 mM sodium phosphate at pH 7.6 with 140 mM urea). Refolding in dilute solutions required detection of the native fluorescence of tryptophan residues using excitation at 248 nm with a KrF excimer laser and collection of emission at 340 nm. The intensity of the fluorescence signal provided information about the environment of the tryptophan residues during the refolding process. Increases in mobility as a consequence of renaturation permitted discrimination of the fully denatured random-coil monomer, a structured monomer intermediate, and a prototrimer which was the precursor of the native tailspike trimer. The folding pathway was temperature-dependent, and CZE could be used to distinguish between productive folding to the native trimer at permissive temperatures and nonproductive aggregation at elevated temperature.

Analysis of Milk and Dairy Products

The analysis of milk and dairy products is important due to the high level of human consumption of these products. The nutritional value of these food sources stems from their significant content of proteins, sugars, vitamins, and fatty acids. Characterization of milk composition is necessary to assure quality and consistency of commercial dairy products. Other areas of research include adulteration, processing stability, storage, genetic selection, and, lately, genetic engineering (by introducing genes that express their products in the milk of mammals, where they can be easily isolated later).

Milk and dairy product polypeptide analyses are frequently performed by HPLC and/or slab gel electrophoresis. HPLC in many cases lacks the selectivity necessary to resolve sample components. Slab gel electrophoresis is labor intensive, and does not provide quantitative information. Despite the drawbacks of slab gel electrophoresis, milk protein nomenclature is largely derived from gel analyses, and therefore the performance of electrophoresis in capillaries as a simplified and yet powerful alternative to gels is very attractive.

Analysis can be performed on whole milk or on whey and casein fractions isolated from milk.

Whey Analysis

Whey is the fraction of milk that remains after removal of caseins by precipitation. A common method to precipitate caseins is to acidify milk to a pH of 4.6 using acetate buffer or hydrochloric acid. β-Lactoglobulins (composed of about 162 amino acids) are the main protein components of whey. There are as many as eight variants of β-lactoglobulin, with variants A and B being the most abundant. Lactoglobulins A and B differ by two amino acids (substitution of Asp for Gly at position 64, and substitution of Val for Ala at position 118), and this provides them with a charge differential that can be exploited for separation in the neutral-to-basic range of the pH scale.

In the authors' lab [192], separation of β-lactoglobulins, BSA and α-lactalbumin is routinely achieved using a 300 mM Na–borate buffer, pH 8.5, and linear polyacrylamide coated capillaries (Figure 14). BSA can be easily resolved from all other whey components after denaturation with SDS and running under sieving conditions. This method does not resolve the β-lactoglobulins [192,193].

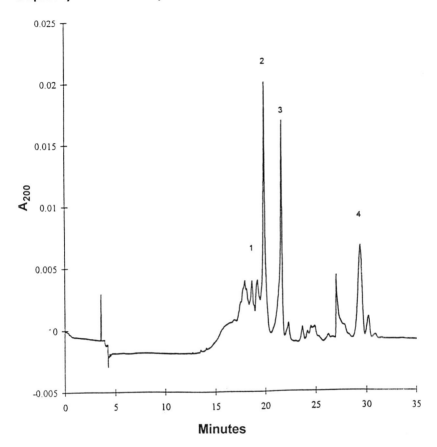

Figure 14 Analysis of whey using a borate buffer at pH 8.5 and a capillary coated with linear polyacrylamide. Peak identities: (1) bovine serum albumin, (2) β-lactoglobulin A, (3) β-lactoglobulin B, (4) α-lactalbumin. Since most proteins in whey have a pI lower than 8, the analysis was performed with injection at the anode.

These proteins plus β-lactoglobulin C have also been analyzed using a 2-(N-morpholino)ethanesulphonic acid (MES) buffer pH 8.0 + 0.1% Tween 20 + 0.1% ethanolamine in uncoated capillaries by Paterson et al. [194]. Although good resolution was achieved with this buffer, it was recognized that the buffer capacity of the solution may be compromised (pKa of MES is 5.3–7.3). Tween and ethanolamine were added to suppress interaction of the polypeptides with the capillary wall. In a later study [195], the same group employed a sample buffer of 10 mM phosphate (pH 7.4) and an analysis buffer of 150 mM borate (pH 8.5) + 0.05% Tween 20 for satisfactory resolution of whey proteins including IgG and BSA.

Another buffer system used with uncoated capillaries is 100 mM borate, pH 8.2 containing 30 mM sodium sulfate [196]. The use of an internal standard and a rinsing protocol that included water, 5 mM sodium hydroxide, water, nitrogen, and run buffer yielded a reproducible method that was used to analyze whey from milk exposed to different treatments, and also milk from several species.

In summary, methods for whey protein analysis employ high pH, coated capillaries, or additives to suppress protein adsorption.

Casein Analysis

The precipitated fraction after milk acidification contains a group of proteins collectively referred to as caseins. This fraction yields a number of bands when analyzed at pH 6.0 using a 100 mM MES buffer in the presence of 8 molar urea [192]. Under these conditions, the milk of different cows under controlled diets produced two sets of patterns, but because the individual casein standards produced more than one peak (especially κ-casein), identification of the sample components was not possible. Other approaches for casein characterization are described below in the section on whole milk analysis.

Peptide Analysis

Several groups have developed CZE methods for separation of enzymatic digests of milk proteins, or to detect naturally occurring peptides (e.g., hormones) and protein degradation products (192,197–199). To assure that all peptides present are electrically charged, electrophoresis is performed at pH extremes. Most commonly, acidic pH is preferred to alkaline pH (at low pH there is little or no EOF, and thus results tend to be more reproducible). Variation of pH between 1.8 and 3.0 greatly influences resolution. When uncoated capillaries are used, adsorption of cationic peptides on the charged silica surface can be significant even at low pH values.

Use of peptide concentration as means to correlate protein integrity and curd texture was established using capillary electrophoresis

[197]. The study also aimed to correlate peptide concentration with bacterial count. Protein standards and peptide hydrolysis products of pure proteins using several proteases were first characterized. A peptide produced by hydrolysis of α_{s1}-casein was selected as a marker because it was present even after extensive plasmin hydrolysis of the parent protein. This peptide was shown to increase in concentration when measured at different time intervals spanning 6 days of milk storage even under refrigeration at 34°F. After 6 days the concentration of the peptide had doubled. At the same time, the hardness of the curd formed by the same milk also decreased, with its lowest point at the sixth day of the analysis. Interestingly, the number of bacterial colonies/milliliter of milk followed a similar pattern as the peptide, reaching a maximum on the sixth day of the experiment.

Whole Milk Analysis
Kanning et al. [200] analyzed milk polypeptides by CZE using a citrate buffer, pH 2.5, containing 6 M urea after dilution of the sample 1:5 with a "reduction buffer" (73 mg trisodium citrate dihydrate + 38 mg DTT + 6 M urea dissolved in 50 ml of water, then titrated to pH 8.0 with NaOH). The separation was carried out in a hydrophilic coated capillary at 45°C. The pH of the buffer was varied between 2.5 and 3.35 to optimize resolution and analysis time (lower pH produced best resolution at the expense of time). This buffer system was shown to produce different patterns for milk of various species (including human). Recio et al. [201] used a similar buffer system (0.32 M citric acid + 20 mM sodium citrate + 6 M urea + 0.05% methylhydroxyethylcellulose, pH 3.0) and a hydrophilic coated capillary to resolve genetic variants (amino acid substitutions and deletions) and posttranslational modifications (phosphorylation) of α and β caseins in reduced bovine milk. This system was also applied to separation of β-lactoglobulin variants in ovine milk, and variations in expression levels of α_{s1} casein in whole caprine milk. Of all methods reviewed, the acidic citrate + urea buffer systems provided the most impressive separations of whole milk proteins.

The same method has recently been applied to the measurement of heat-denatured serum proteins in heat-treated milk by analysis of the casein fraction obtained by isoelectric precipitation [202].

Pretreatment of the coated capillary with bovine serum albumin was found to reduce the loss of BSA by adsorption during analytical separations.

All major milk polypeptides can also be analyzed using a 100 mM phosphate buffer, pH 2.5, containing 8 M urea [192]. The sample is simply prepared by adding solid 8 M urea to the milk, incubating for 5 min at 40°C (to dissolve the urea), and centrifuging for 5 min to remove fat and precipitates that may clog the capillary. The remaining clear (yellowish) solution can be injected directly into the column after adjusting the pH with a 1:10 dilution of the run buffer (dilution of the run buffer is recommended to achieve zone sharpening, thus increasing resolution and sensitivity). Using this method, electropherograms with different patterns were obtained for bovine and human milk, and for milk samples obtained from different bovines which presumably reflected genotypic differences (Figure 15).

Whole milk can also be analyzed using a 250 mM borate buffer, pH 10.0 in uncoated capillaries [203]. Using this system, β-casein and α-lactalbumin were resolved from all other sample components, but casein and β-lactoglobulin migrated very close to each other (without complete resolution) in fresh nonfat milk. A dramatic difference was observed when analyzing powdered milk: β-lactoglobulin was totally absent, and both caseins appeared as broad peaks, possibly due to denaturation during the drying process.

Metalloproteins

Capillary zone electrophoresis has been widely applied to the separation of metalloproteins and metal-binding proteins, and this subject has been recently reviewed by Richards and Beattie [204]. Applications include determination of purity, identification of isoforms and genetic variants, structural studies, metal-binding studies, stability studies, and determination of enzymatic activity.

The oxidation state of protein-bound metal ions can affect migration time, and CZE has been used to identity redox changes in proteins. In a study on the influence of capillary temperature in CZE of proteins, Rush et al. [205] noted that electrophoresis of the heme-containing protein myoglobin at elevated temperature under constant current operation resulted in the appearance of a second slower-

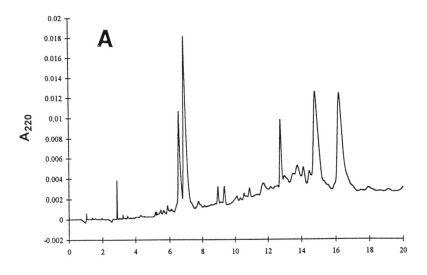

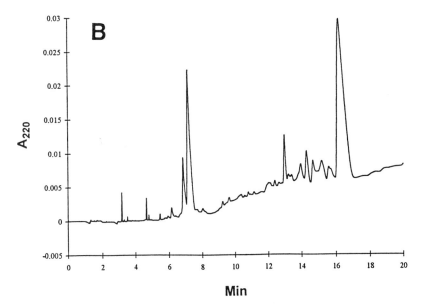

Figure 15 Analysis of bovine milk by CZE at pH 6.0. The two panels (A and B) represent milk from two individual cows fed the same controlled diet. All analyses were performed using 0.1 M MES buffer containing 8 M urea in a linear polyacrylamide-coated capillary, injection at the anode, and detection at 220 nm.

migrating species. Further investigation demonstrated that this was not due to protein conformational change but to reduction of the heme-associated iron from the ferric to ferrous state. It was speculated that on-column reduction might be due to a reducing impurity in the electrophoresis buffer or to autoreduction by residues in the polypeptide; in any case, this phenomenon should be considered in the development of separation methods for metalloproteins. Sun and Hartwick [128] employed a multipoint detection scheme to monitor the conversion of myoglobin Fe^{3+} to Fe^{2+} during the separation, and used this system to determine the effect of temperature on the myoglobin degradation rate constant. Lee and Yeung [206] analyzed the proteins in a population of 29 human red blood cells by introducing single erythrocytes into a capillary; each cell was lysed after injection, proteins were separated electrophoretically and detected by native laser-induced fluorescence. Variations in the relative ratios of hemoglobin A and methemoglobin were observed and attributed to variations in in vitro oxidation reflecting the age distribution of the sampled erythrocytes.

Effect of metal binding on electrophoretic mobility for three metal-binding proteins was investigated by Kajiwara [207]. Using Tris-tricine or Tris-glycine buffers, it was observed that addition of calcium ions to the electrophoresis buffer caused an increase in the mobility of the calcium-binding proteins calmodulin and parvalbumin. This change in mobility was attributed mainly to the change in net charge rather than changes in Stokes radii of the proteins. Carbonic anhydrase, a zinc-binding protein, did not display a shift in migration time with a zinc-containing electrophoresis buffer, perhaps due to weak affinity of the active site for Zn^{2+} ions. Thermolysin is a metalloprotease which binds four Ca^{2+} ions for conformational stability and one Zn^{2+} ion at the active site. In the presence of EDTA thermolysin migrated as a broad peak suggesting conformational instability; addition of Zn ions reduced mobility but did not improve peak shape. Upon addition of calcium ions or both calcium and zinc ions thermolysin appeared as a very sharp peak with increased mobility. Metal-dependent migration shifts were also observed for calmodulin, parvalbumin, and thermolysin using MEKC, indicating that metal binding reduces the hydrophobic character of the protein.

Metallothioneins are low-molecular-weight, cysteine-rich metal-binding proteins which in many species exist in a variety of isoforms due to genetic polymorphism. Application of CZE to the separation of metallothionein isoforms was first described by Beattie et al. [208] using a Tris-HCl buffer at pH 9.1 in an uncoated capillary. These conditions could easily resolve MT-1 and MT-2, the two major metallothionein classes, and the method was used for determination of metallothioneins in liver extracts from Zn-induced rats (Figure 16). Removal of interfering protein by precipitation or acetonitrile was necessary for analysis of tissue samples. Liu et al. [209] described a Tris-sodium tetraborate buffer system for separation of MT-1 and MT-2 using an uncoated capillary, but the method was unsuit-

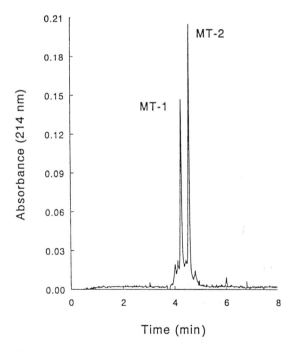

Figure 16 Capillary zone electropherogram of rabbit liver metallothioneins (0.50 mg/ml). Analysis parameters: capillary, 94 cm (74 cm effective length) uncoated; injection, 1 s at 30 kV; buffer, 50 mM Tris-HCL (pH 9.1), run voltage, 30 kV. (Reproduced from Ref. 208 with permission.)

able for metallothionein analysis in tissue samples due to protein adsorption to the capillary. Virtanen et al. [210] used Tris-borate at high concentration (110 mM each, pH 6.9) for the separation of rabbit liver and horse kidney metallothioneins in uncoated capillaries.

Separations of metallothionein isoforms under a variety of CZE conditions has been reported by Richards et al. [211–213]. In addition to resolution of MT-1 and MT-2, separations also revealed heterogeneity within the major groups. The authors compared separations achieved using uncoated fused silica capillaries, capillaries dynamically coated with a polyamine polymer, and capillaries covalently coated with polyacrylamide. Separations in uncoated capillaries were performed under alkaline conditions using a borate buffer and neutral or acidic conditions using phosphate buffers. Use of borate buffer at pH 8.4 separated MT-1 and MT-2 but did not resolve isoforms within the major classes. Use of phosphate buffer at pH 7 reduced EOF and allowed resolution of additional isoforms. Phosphate buffer also provided better sensitivity due to its reduced absorbance in the low UV; this was advantageous due to the lack of aromatic residues in metallothioneins. Richards et al. [214] used similar conditions to resolve MT-1 and MT-2 (isolated from zinc-induced mouse liver) from recombinant mouse MT-3 (originally identified in brain tissue). Under acidic conditions, zinc and cadmium ions dissociate from the polypeptides, and use of phosphate buffer at pH 2.0 permitted resolution of apothionein isoforms. Absence of metals was confirmed by use of UV scanning detection which yielded spectra characteristic of the metal-free proteins. Scanning detection could also be used to distinguish the type of metal bound to the metallothioneins under neutral conditions. The use of very high ionic strength buffers (0.5 M) provided improved resolution and rapid analysis due to reduction of EOF, but required the use of small-bore (20 μm i.d.) capillaries to minimize Joule heat. Further improvement in separations of metallothionein isoforms was obtained using the polyamine-coated capillary with reversed EOF at pH 7, but resolution was highly dependent on the composition of the electrophoresis buffer; optimal separations were obtained using 100 mM sodium phosphate buffer at pH 7.0. The neutral polyacrylamide-coated capillary, which exhibited neglible EOF, permitted separations to be performed with either normal or reversed polarity over a wide pH range to achieve additional selectivity, and provided improved resolution at pH 7 compared to either uncoated or polyamine-coated capillaries. Using a polyacrylamide-coated capillary with a HEPES buffer (pH 7.4), Minami et al. [215,216] observed a reduction in MT-1 and MT-2 peak areas upon addition of cadmium to purified rat liver metallothionein or crude mouse liver metallothioneins. Loss of metallothionein response was dependent upon the duration of incubation with Cd and with Cd concentration. This loss was not observed with addition of zinc or sodium; Cd-dependent peak loss could be reversed by the addition of ethylene glycol bis-(β-aminoethyl ether) $N,N,N'N'$-tetraacetic acid.

A class of metallothioneins found in plants, algae, and fungi are not direct gene products and are composed of repeating units of γ-glutamylcysteine. Torres et al. [217] analyzed these class III metallothioneins in crude extracts of Cd-induced mircoalgal cells. Use of acidic condtions with low but measurable EOF and reversed polarity allowed the separation of six MT III species under conditions where cathodically migrating neutral and cationic interfering components of the extracts were not detected.

Beattie and Richards [218] employed MECC for successful separation of metallothionein isoforms from a variety of species using a borate buffer containing 75–100 mM SDS. Interestingly, metallothioneins typically exhibit broad bands in SDS-PAGE due to formation of inter- and intramolecular linkages via cysteine groups, but produced sharp peaks in the MECC separations. These authors concluded that CZE in uncoated capillaries with alkaline buffers can be used for analysis of metallothioneins which exhibit little heterogeneity beyond the two major classes. For metallothioneins that exhibit significant microheterogeneity or for analysis of metallothioneins in complex sample matrices, other approaches such as CZE under acidic conditions, CZE with coated capillaries, or use of MECC may be preferred.

Application of capillary zone electrophoresis for characterization of ferritin subunit composition and associated ferritins was reported by Zhao et al. [219]. Ferritins are multisubunit proteins functioning in the concentration and storage of iron. The spherical holoferritin complex consists of 24 subunits organized around a mineral core containing up to 4500 iron atoms and variable amounts of inorganic phosphate. These authors have used CZE in coated capillaries with a variety of buffer systems to study the purity of ferritin, the subunit composition of ferritin preparations, and assembly of the ferritin complex from individual subunits. The latter study derived from the fact that ferritin exists as subunits at low pH but as the native complex at physiological pH. Results indicated that apo, holo, and reconstituted ferritins have the same mobility, suggesting that the iron core does not change the overall charge density. However, during iron deposition in the assembly process, significant mobility shifts were observed, suggesting that CE may be a useful tool for studying the process of core formation in both mammalian and bacterial ferritins. One caution in these studies was the observation that migration behavior in the presence of variable amounts of iron was strongly affected by the buffer composition, indicating that the electrophoretic conditions could markedly affect metal binding.

Application of CZE to characterization of a fibrinolytic metalloproteinase was described by Markland et al. [220]. Snake venom fibrolase, a 23 kDa zinc-containing enzyme purified from southern copperhead venom, was resolved into two isoforms using an uncoated capillary with 100 mM Tris-HCl at pH 7.6. Crystalline fibrolase was also resolved into two isoforms, while recombinant fibrolase could be resolved into three components. The authors were also able to use CZE in the presence of urea to monitor the inactivation of the metalloenzyme by incubation with EDTA, since the apoenzyme was well separated from the Zn-associated protein.

Improved sensitivity for detection of metal-binding proteins using an on-line preconcentration scheme was described by Cai and El Rassi [221]. A preconcentration capillary containing a metal chelating coating was used in tandem with the separation capillary. The preconcentration capillary was prepared by first etching the interior of the silica tube to increase the surface area, then attaching a covalent hydrophilic coating with activated epoxide groups. Finally, iminodiacetic acid (IDA) was attached to the activated coating, and this moiety, after complexation with zinc ion, served as an affinity ligand for selectively concentrating metal-binding proteins. The preconcentration column was directly coupled to the separation capillary with a PTFE union; the separation capillary was also coated with a covalently linked polyether to minimize protein adsorption. In the initial binding step, a dilute solution of the metal-binding protein was introduced into the capillary by gravity; interaction of certain amino acid side chains (primarily histidine and cysteine) with the IDA affinity group enabled concentration of the protein on the surface. In a subsequent debinding step, a strongly competing ligand (EDTA) was introduced to elute the concentrated protein into the separation capillary as a narrow band. This system was able to increase detection sensitivity 25-fold for carbonic anhydrase; optimum binding was obtained at neutral pH while optimum debinding was achieved with 50 mM EDTA at pH 3.5.

Glycoproteins

Glycoproteins often exist as multiple glycoforms sharing a common amino acid primary sequence but differing in the number, location, and structure of carbohydrate groups attached to the polypeptide chain. The importance of glycosylation patterns in the biological activity of glycoproteins has generated strong interest in methods for separation of glycoforms, particular in the case of therapeutic glycoproteins. Variation in the number of sialic acid residues confers charge microheterogeneity to glycoproteins, and gel electrophoresis and isoelectric focusing have been used successfully for their characterization. Therefore, CZE and CIEF are obvious candidates for automated analysis of protein glycoforms. A typical strategy for optimizing CZE separations of glycoforms starts with determination of conditions which yield best resolution of glycoforms, followed by enzymatic cleavage (e.g., with neuraminidase) of carbohydrate moieties. The disappearance of peaks in the electropherogram following enzyme treatment confirms their identity as glycoforms.

Separation of protein glycoforms was first described by Kilár and Hjertén [222], who used CZE with a Tris-borate + EDTA (pH 8.4) buffer and a coated capillary to separate the di-, tri-, tetra-, penta- and hexasialo isoforms of iron-free human serum transferrin. Yim [223] used CZE to resolve glycoforms of recombinant tissue plasminogen activator (rtPA), a 60 kDa glycoprotein containing complex N-linked oligosaccharides attached to the polypeptide chain at two (type II) or three (type I) sites. Using an ammonium phosphate buffer at pH 4.6 containing 0.01% reduced Triton X-100 + 0.2M ε-aminocaproic acid (added to stabilize solubility of the protein) in a linear polyacrylamide-coated capillary, approximately 15 glycoforms were partially resolved. Recently Thorne et al. [224] at the same institution expanded this study and found that other ω-amino carboxylic acids were less effective than EACA in achieving glycoform resolution, and that the addition of the Tween 80 surfactant was necessary to obtain good recovery.

Erythropoeitin (EPO) is an acidic (pI 4.5–5.0) 35 kDa glycoprotein with three N-linked and one O-linked polysaccharide chains comprising 40% of the protein mass. Tran et al. [225] investigated the effects of buffer pH and organic modifiers and achieved partial resolution of five glycoforms using an acetate–phosphate buffer at pH 4.0. However, column equilibration times of up to 11 h were required for good reproducibility. Watson and Yao [226] were able to achieve complete separation of six EPO glycoforms by adding 1,4-diaminobutane (DAB) and 7 M urea to a tricine–NaCl buffer at pH 6.2. The alkylamine additive reduced EOF and allowed glycoforms to be resolved in order of increasing sialic acid content. Bietlot and Girard [227] developed an assay for recombinant human EPO (rhEPO) in formulations containing high concentrations of human serum albumin as an excipient. Using an amine-coated capillary with reversed EOF, the separation of rhEPO into several glycoforms was acheived with a 200 mM phosphate buffer at pH 4. They found that the HSA excipient comigrated with EPO under these conditions, but determined that addition of 1 mM nickel chloride permitted resolution of HSA from rhEPO with no degradation of the EPO glycoform separation pattern.

Landers et al. [52,53] used DAB as well as hexa- and decamethonium salts as additives for separation of glycoforms of ovalbumin, pepsin, and human chorionic gonadotropin. Legaz aand Pedrosa [65] evaluated a series of alkylamines and polyamines for resolution of ovalbumin glycoforms and achieved optimal separations with 25 mM phosphate (pH 9.0) containing 0.87 mM spermidine or 0.14 mM spermine.

Morbeck et al. [228] used a 25 mM borate buffer (pH 8.8) containing 5 mM 1,3-diaminopropane to resolve up to 8 glycoforms of human chorionic gonadotropin, a 38 kDa glycoprotein consisting of an α subunit with two N-linked glycosylation sites and a β subunit with the two *N*-linked sites plus four *O*-linked glycosylation sites.

Bovine pancreatic ribonuclease is a mixture of unglycosylated Rnase A and a family of five glycoforms bearing variable-length oligomannoses at a single *N*-linked site (Rnase B Man-5 to Man-9). Using a buffer consisting of 20 mM sodium phosphate + 5 mM sodium tetraborate + 50 mM SDS (pH 7.2), Rudd et al. [229] were able to separate all five glycoforms, enabling the use of capillary electrophoresis as a tool in determining the importance of glycosylation in the stability and functional properties of the enzyme.

Recombinant human bone morphogenetic protein (rhBMP) is a basic dimeric protein consisting of two identical 114-residue subunits with single glycosylation sites at Asn56 carrying two N-acetylglucosamines and 5–9 mannose units. Using a pH 2.5 phosphate buffer with coated capillaries, Yim et al. [230] were able to resolve the 15 possible glycoforms into 9 peaks, each of which differed by only one mannose unit (Figure 17). The authors demonstrated that mobilities of the glycoforms decreased in proportion to the number of mannose units present, and concluded that frictional drag of the large mannose residues had a strong effect on migration.

Chen (231) investigated the effect of several separation parameters on resolution of ovalbumin glycoforms, including buffer composition and pH, use of buffer additives, and use of capillary coatings. Best results were achieved using polyacrylamide-coated capillaries and a borate buffer prepared with aminoalcohols such as Tris or aminomethylpropanol, which resolved ovalbumin into over 20 components.

Cereal Proteins

The quality of food products made from wheat and other cereals is often dependent upon the type and content of storage proteins in the grain endosperm. Grain quality differences among different varieties has been characterized by extraction of the endosperm proteins and comparison of the separation patterns obtained with native PAGE under acidic conditions, SDS-PAGE, and reversed-phase HPLC. Several groups have investigated CZE as an alternative method for characterization of cereal proteins.

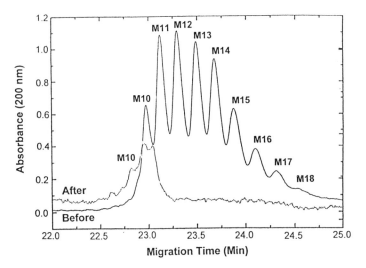

Figure 17 Overlay of the CZE profiles of intact (solid trace) and α(1,2)-mannosidase digested (dotted trace) recombinant human bone morphogenetic protein. The labeled peaks indicate individual glycoforms of the dimeric protein which differ by one mannose residue at the single glycosylation site on each monomer. Glycoforms which possess the same number of sugar residues were not resolved under these conditions, resulting in separation of the 15 glycoforms into 9 peaks. (Reproduced from Ref. 230 with permission.)

Bietz [232,233] compared CZE in uncoated capillaries under basic and acidic conditions for separation of gliadins extracted with 30% ethanol from wheat seeds and flour. Basic conditions (60 mM borate buffer at pH 9 containing 20% acetonitrile and 1% SDS) at 40°C yielded resolution comparable to reversed-phase HPLC, but reproducibility was poor without extensive capillary conditioning and between-run wash protocols. Use of acidic conditions (100 mM sodium phosphate buffer at pH 2.5 containing hydroxypropylmethylcellulose, HPMC) provided improved reproducibility and significantly higher resolution. The patterns obtained under these conditions for extracts from six wheat varieties indicated CZE could be successfully used to differentiate genetically related varieties.

Werner et al. [234] employed CZE under acidic conditions for separation of gliadens extracted from wheat with 70% ethanol. In this study, the capillary was dynamically coated with a cationic polymer which reversed the direction of EOF and minimized protein adsorption. Four gliaden fractions (α, β, γ, ω) were purified by gel filtration and analyzed by acidic PAGE and CZE; the four fractions exhibited the same relative mobility in both techniques. Application of this method to separation of 12 cultivars yielded distinguishable patterns for each cultivar.

Lookhart and Bean [235] developed a rapid method for wheat cultivar differentiation using an acidic phosphate buffer containing HPMC with 20 μm uncoated capillaries at 45°C. The migration order of the major endosperm protein classes (albumins, globulins, gliadins, and glutenins) was established, and the migration order of the gliadin subclasses was compared with acid-PAGE, SDS-PAGE, and HPLC [236]. This method was able to differentiate closely related cultivars which were not differentiable by acidic PAGE. These authors [237] used the same analysis conditions to characterize endosperm proteins from other grains as well. Avenins were extracted from oats using 70% ethanol, and prolamins were extracted from rice using 60% n-propanol. CZE separation patterns could be used to differentiate U.S. oat cultivars and to distinguish both U.S. and Philippine rice cultivars. Differences in CZE protein profiles were observed for cultivars which exhibited identical profiles with reversed-phase HPLC or acidic PAGE. This method was later refined [238] by using a single 1 M phosphoric acid wash between injections (which improved reproducibility) and by addition of 20% acetonitrile to the electrophoresis buffer (which improved resolution). These optimized conditions were applied to the identification of proteins associated with the presence of wheat–rye chromosomal translocations in several wheat cultivars [239]. The novel translocation-associated proteins migrated in the ω-gliadin region, as expected for proteins (secalins) of rye origin (Figure 18). Capillary zone electrophoresis has also been used to characterize soluble proteins remaining after coagulation of wheat albumin and globulin fractions at elevated temperatures [240].

Wong et al. [241] resolved the 7S-rich (conglycinin) and 11S-rich (glycinin) fractions of soy protein using an uncoated capillary with a 20 mM borate buffer at pH 8.5, and used these conditions to monitor protein hydrolysis with several proteases.

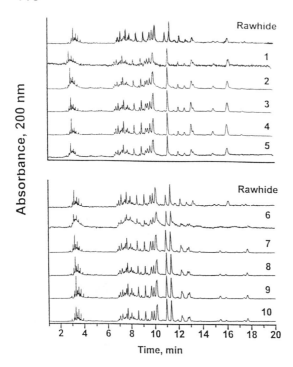

Figure 18 Comparison of CZE protein profiles of sister lines of the wheat cultivar Rawhide which do carry (traces 1–5) or do not carry (traces 6–10) the 1Bl.1RS rye protein chromosomal translocations. Protein peaks characteristic of the translocation migrate as a doublet around 13 min. (Reproduced from Ref. 239 with permission.)

Collagens

Collagens comprise a family of structurally related connective tissue proteins. The various collagen types are typically isolated and characterized by liquid chromatography or SDS-PAGE, either as intact chains or as peptide cleavage products [242]. Analysis of collagens by CE has been a challenge due to their high molecular weight, poor solubility, limited UV absorbance, and their tendency to interact with the capillary wall. Collagen separation by CE under various conditions has been extensively investigated by Deyl's group. Using a 2.5

mM sodium tetraborate buffer at pH 9.2 with an uncoated capillary, collagen chains (α, α_1, α_2, α_3, β, β_{11}, β_{12}, γ) from collagen types I, II, V, IX, and XI were resolved; CNBr peptides were resolved using the same buffer at pH 10.2 [243]. Adsorption of collagen to the capillary wall was a problem with this system. In a later publication [244] collagen type I chains (α_1, α_2, β_{11}, β_{12}, γ) were resolved using a dynamic sieving system employing linear polyacrylamide and SDS (see Chapter 7).

The separation of CNBr fragments of collagen by CZE were compared under alkaline and acidic conditions [245,246]. Separation order using 2.5 M sodium tetraborate buffer at pH 9.5 was dependent on mass/charge ratio of the fragments [245], while separation order using 100 mM sodium phosphate at pH 2.5 was similar to that of the same fragments separated by reversed phase HPLC, e.g., in order of increasing hydrophobicity or molecular mass [246]. The authors attributed this to hydrophobic interaction of the fragments with the capillary wall, an interpretation supported by reduced migration time and resolution of the same fragments when analyzed using the same low pH conditions with a capillary coated with hyrophilic linear polyacrylamide. This study was expanded in a later report to evaluate the effects of buffer ionic strength and capillary temperature [247]. Using optimized conditions (25 mM phosphate buffer, pH 2.5, and 50°C), the method was used to determine the proportion the major fiber-forming collagens (types I, III, and V) in a variety of tissue samples by quantitation of the peptides released by CNBr digestion (Figure 19). Similar conditions were used to differentiate the collagen structure of young and aged rats by the presence of high molecular-mass CNBr peptides [248].

Purification and Process Monitoring

Because of the speed and high resolution of CZE separations, and because information about the composition of complex protein samples can be obtained from extremely small volumes, capillary electrophoresis is increasingly being used to assess protein purity in multistep purification protocols on the laboratory, pilot plant, and process scales. Similarly, it is being considered as a candidate for monitoring fermentations. McNerney et al. [249] described

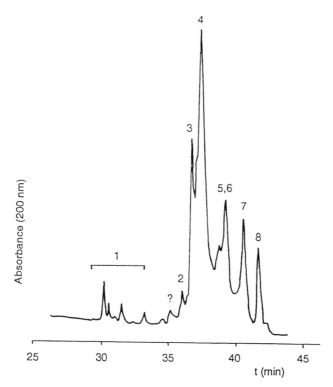

Figure 19 Separation of CNBr peptides derived from type I collagen by capillary zone electrophoresis. Analysis conditions: capillary, 50 cm × 50 μm uncoated; injection, 8 kV for 4 sec; buffer, 0.1 M phosphate, pH 2.5; run conditions, 8 kV with positive to negative polarity. (Reproduced from Ref. 246.)

separation of recombinant human growth hormone (rhGH) and its variants from very crude mixtures of *E. coli* cell extracts using CZE in a phosphate-deactivated capillary. The 18 h deactivation procedure included washing the capillary with 0.1 M nitric acid and 0.1 M sodium hydroxide to remove contaminants prior to conditioning with the run buffer (250 mM sodium phosphate + 1% propylene glycol). This method allowed separation of a variety of rhGH variants including deamidated and dideamidated rhGH, desPhe- and desPhePro rhGH, and 2-chain rhGH, and could be used to detect changes in fermentation conditions which affected rhGH production. Between-run washing with 3 M guanidine HCl + 0.2 M sodium phosphate was required to remove adsorbed contaminants. The phosphate-deactivated column provided superior resolution and reproducibility compared to bare fused-silica or PVA-coated capillaries.

Purification of murine antiheparin monoclonal antibody produced in cell culture was monitored by Malsch et al. [250] using a CZE with a borate/boric acid buffer (pH 9.0) in an uncoated capillary.

Kundu et al. [251] used MEKC conditions to assess the purity of two recombinant proteins, a cytomegalovirus-CMP-KDO synthetase fusion protein expressed in *E. coli* and a hepatitis C viral protein expressed in CHO cells. Proteins were prepared in a 10 mM Tris–1% SDS buffer (pH 8.5) and analyzed in a 10 mM borate–100 mM SDS buffer (pH 9.5) in uncoated capillaries. The level of impurities, which varied with the method of protein production, was found to agree within ±5% with results obtained by densitometric scanning of SDS–PAGE gels of the same materials.

Serum and Urinary Proteins

Automated analysis and quantitative information are key advantages in the clinical laboratory, and therefore capillary electophoresis is being intensively evaluated by many clinical research groups as a replacment for labor-intensive or semiquantitive procedures routinely used in clinical diagnostics. Characterization of the protein profiles in serum and urine are two high-volume tests which are candidates for transfer from conventional gel electrophoresis to CZE. Application of CE to analysis of proteins in clinical samples is reviewed by Jellum [252] and Jenkins and Guerin [253].

Serum Proteins
Agarose gel electrophoresis is widely used in clinical laboratories as a screening tool for detecting protein abnormalities in serum and other biological fluids. Clinical chemists have shown interest in using CZE for quantitative automated analysis of proteins in clinical samples. For simplicity and robustness, most approaches use uncoated fused silica capillaries operated at a pH above the pI of analyte proteins to reduce wall interactions. Under these conditions, the high rate of EOF

Patient Data
Sample ID: KH25
Patient ID:
Name:
Sex:
DOB:
Physician:

Analysis Data
Analysis Performed: 09/10/97 12:40
Injection Number: 1855
Vial Number: 13
Run Number: 172
Report Generated: 09/10/97 12:44
Operator ID:
Supervisor ID:

Sample Data
Total Protein: 7.4 g/dL
Comments:

Peak Name	Area	Area%	Reference Range (%)	Conc (g/dL)	Reference Range (g/dL)
Albumin	6069	58.6	51.3 - 66.6	4.33	3.60 - 5.03
Alpha1	423	4.1	3.8 - 6.6	0.30	0.27 - 0.48
Alpha2	893	8.6	6.8 - 13.0	0.64	0.47 - 0.92
Beta	1273	12.3	8.8 - 15.7	0.91	0.61 - 1.22
Gamma	1707	16.5	9.4 - 19.5	1.22	0.65 - 1.54

Total area: 10365

A/G ratio: 1.41

Analysis comments:

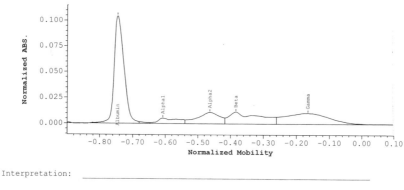

Interpretation: _____
                 _____

Reviewer: _____   Date: _____

Figure 20 Separation of serum proteins from a normal patient using a commercial CE serum protein analysis kit from Bio-Rad Laboratories.

results in migration of proteins in order of decreasing pI (e.g., in order of increasing electrophoretic mobility at alkaline pH). The migration order in the resulting electropherogram is the mirror image of the densitometer trace obtained from an agarose protein electrophoresis gel. Protein separations of normal and abnormal sera using

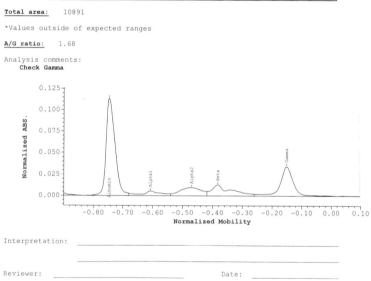

Patient Data
Sample ID: KH24
Patient ID:
Name:
Sex:
DOB:
Physician:

Analysis Data
Analysis Performed: 09/10/97 12:32
Injection Number: 1854
Vial Number: 12
Run Number: 172
Report Generated: 09/10/97 12:36
Operator ID:
Supervisor ID:

Sample Data
Total Protein: 7.8 g/dL
Comments:

Peak Name	Area	Area%	Reference Range (%)	Conc (g/dL)	Reference Range (g/dL)
Albumin	6824	62.7	51.3 - 66.6	4.89	3.60 - 5.03
Alpha1	409	3.8*	3.8 - 6.6	0.29	0.27 - 0.48
Alpha2	828	7.6	6.8 - 13.0	0.59	0.47 - 0.92
Beta	1017	9.3	8.8 - 15.7	0.73	0.61 - 1.22
Gamma	1811	16.6	9.4 - 19.5	1.30	0.65 - 1.54

Total area: 10891

*Values outside of expected ranges

A/G ratio: 1.68

Analysis comments:
    Check Gamma

Interpretation: _____

                _____

Reviewer: _____   Date: _____

Figure 21 Separation of serum proteins from a patient exhibiting an elevated peak in the γ-region. Analysis was performed using a commercial CE serum protein analysis kit from Bio-Rad Laboratories.

a commercial CE system with 5-band and high-resolution buffer kits are shown in Figures 20 to 23.

Electrophoretic separation of serum proteins in a glass capillary was first demonstrated by Hjertén in 1967 [254] using a 3 mm tube containing a Tris-HCl buffer (pH 8.7). In 1983, Jorgenson and Lukacs [255] used a 75 μm i.d. surface-modified capillary to resolve human serum into several peaks with a 0.1 M Tris-HCl (pH 8.5) electrolyte.

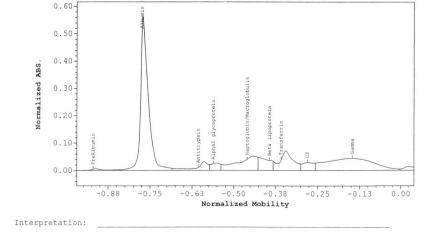

Patient Data
Sample ID: KH25
Patient ID:
Name:
Sex:
DOB:
Physician:
Total Protein: 74.00 g/L

Analysis Data
Analysis Performed: 09/09/97 12:39
Injection Number: 875R
Vial Number: 8
Run Number: 72
Report Generated: 09/10/97 13:52
Operator ID: LAP
Supervisor ID:

Protein Fractions	Conc (g/L)	Reference Range (g/L)
PreAlbumin	0.37	0.10 - 0.60
Albumin	40.89	36.00 - 50.90
Antitrypsin	2.14	1.80 - 3.60
Alpha1 glycoprotein	1.39	0.70 - 1.40
Haptoglobin/Macroglobulin	6.10	4.10 - 8.20
Beta lipoprotein	3.10	1.90 - 4.10
Transferrin	5.53	3.90 - 6.60
C3	1.95	1.10 - 2.30
Gamma	12.53	6.80 - 15.60

Total Area: 44107   **A/G ratio:**   1.25

Analysis comments:
  Check Gamma
PreAlbumin =   0.5 %   Alpha2 Zone =   8.2%
Albumin Zone =   55.3 %   Beta Zone =   14.3%
Alpha1 Zone =   4.8%   Gamma Zone =   16.9%

Interpretation: _____

Reviewer: _____ Date: _____

Figure 22   Separation of serum proteins from a normal patient performed using a commercial high-resolution CE serum protein analysis kit from Bio-Rad Laboratories.

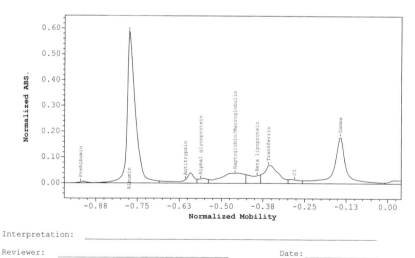

Patient Data

Sample ID:	KH24
Patient ID:	
Name:	
Sex:	
DOB:	
Physician:	
Total Protein:	78.00 g/L

Analysis Data

Analysis Performed:	09/09/97 12:28
Injection Number:	874R
Vial Number:	7
Run Number:	72
Report Generated:	09/10/97 13:51
Operator ID:	LAP
Supervisor ID:	

Protein Fractions	Conc (g/L)	Reference Range (g/L)
PreAlbumin	0.34	0.10 - 0.60
Albumin	46.29	36.00 - 50.90
Antitrypsin	2.44	1.80 - 3.60
Alpha1 glycoprotein	0.98	0.70 - 1.40
Haptoglobin/Macroglobulin	5.53	4.10 - 8.20
Beta lipoprotein	2.33	1.90 - 4.10
Transferrin	6.02	3.90 - 6.60
C3	1.00*	1.10 - 2.30
Gamma	13.07	6.80 - 15.60

*Values outside of expected ranges

Total Area: 45744   **A/G ratio:** 1.48

Analysis comments:
  Check Gamma

PreAlbumin =	0.4 %	Alpha2 Zone =	7.1%
Albumin Zone =	59.3 %	Beta Zone =	12.0%
Alpha1 Zone =	4.4%	Gamma Zone =	16.8%

Interpretation: _____

Reviewer: _____   Date: _____

Figure 23 Separation of serum proteins from a patient exhibiting an elevated peak in the γ-region. Analysis was performed using a commercial high resolution CE serum protein analysis kit from Bio-Rad Laboratories.

Chen and co-workers were the first to develop CE methodology for routine clinical analysis of serum proteins [156,256,257]. They used a proprietary borate-based buffer at pH 10 and an uncoated 20-25 μm i.d. × 25 cm capillary operated at 20 kV to resolve human serum proteins into five bands (beta, alpha2, alpha1, albumin, and prealbumin) in less than 100 s. Increasing buffer ionic strength was found to improve resolution, presumably by decreasing band broadening due to protein–protein interactions. Under these conditions, compliments, transferrin, β-liproteins, haptoglobin, α_2-macroglobulin, α_1-antitrypsin, and α_1-lipotroteins appeared as discreet peaks. Capillary washes with water and base improved migration time reproducibility. Capillary electrophoresis profiles obtained for normal and myeloma patient serum samples were similar to those obtained with agarose gel electrophoresis. The same system was used to analyze urine and CSF proteins; in the latter case, nonprotein sample components were removed by dialysis. Application of an automated system based on this separation chemistry for assessment of the stability of serum protein profiles in blood bank samples has been reported by Jellum et al. [258,259].

Kim et al. [260] separated serum proteins into five bands using 30 mM sodium borate at pH 9.4 and an uncoated 50 μm i.d. × 100 cm capillary operated at 30 kV. Results for the five protein fractions were compared with those obtained by agarose gel electrophoresis, and good correlation was observed for the major albumin and γ fractions while correlation for the minor components was poorer. Serum protein separations using similar conditions were reported by Lehmann et al. [261]. Reif et al. [262] described serum protein analysis using 40 mM borate at pH 10.6. The use of a short 50 μm i.d. × 27 cm capillary enabled the five protein bands to be resolved in 90 s. To regenerate capillaries, they were purged with base, water, and nitrogen before replenishing the borate analysis buffer; detergent washes were included in the regeneration cycle if necessary to remove strongly adsorbed proteins.

Jenkins et al. [263,264] analyzed proteins in serum by CZE using a 50 mM borate buffer (pH 9.7) with 1 mM calcium lactate. The calcium lactate was added to enhance the separation of the beta band components transferrin and C_3. In this study, 1000 clinical samples were analyzed both by CZE and high-resolution agarose gel electrophoresis (HRAGE). For the 362 samples containing one or more monoclonal protein bands, comparison of CZE and HRAGE values for monoclonal levels from 1 to 71 g/l yielded a correlation coefficient of 0.96. Statistical analysis indicated that HRAGE gave slightly greater monoclonal levels than CZE.

Dolnik [265] evaluated a series of electrolytes for serum protein analysis. Sodium, Tris, and N-methyl-D-glucamine (MGA) were investigated as counterions in combination with tricine, asparagine, boric acid, glycine, γ-aminobutyric acid, ε-aminocaproic acid (EACA), and lauric acid as coions. Best results were obtained with a buffer composition of 0.1 M MGA + 0.1 M EACA, which resolved normal serum into 10 protein zones.

Clark et al. [266] compared CZE using a 150 mM borate buffer in uncoated capillaries with agarose gel electrophoresis for analysis of over 300 patient serum samples. The 120-s CZE separation provided comparable information to AGE, and could be used to interpret samples identified as "point of application" artifacts in AGE. The authors noted that direct on-tube UV detection of separated proteins eliminated detection anomalies in AGE associated with variable dye binding, and therefore should more reliably reflect true quantitative variations in protein content.

Using a 20 mM borate buffer at pH 10, Lehman et al. [261,267] compared the five-band profiles obtained with CZE and cellulose acetate membrane electrophoresis (CAME) from 102 patient samples. The α_1-, α_2-, β-, and γ-globulin results for the two techniques agreed well while albumin results were in poorer agreement, perhaps due to variable CAME staining efficiencies at low protein concentration.

A different approach to separation of serum proteins was described by Andrieux et al. [268] who employed a 25 mM phosphate–25 mM borate buffer (pH 8.5) containing 50 mM SDS. Not surprisingly, migration order was different from CZE under native conditions, and glycoproteins tended to migrate faster than nonglycosylated proteins.

Application of capillary zone electrophoresis to analysis of serum cryoglobulins has been described by Shihabi [269]. These are immunoglobulins which precipitate reversibly from serum at cold temperatures, producing a variety of clinical disorders. They are typically classified as type I (monoclonal), type II (mixed polyclonal with a monoclonal component), or type III (mixed polyclonal).

Cryoglobulins are identified by comparing the protein profile of the precipitate remaining after cold centrifugation of a serum sample with that of the uncentrifuged control. Advantages over traditional agarose gel electrophoresis include reduction in required sample volumes, reduction in sample preparation steps, direct quantification, and increased sensitivity to monoclonal components. A limitation of the CZE method is its inability to characterize the exact immunological composition of the cryoglobulins.

Hemoglobin Variants

Hemoglobin variants arise from genetic changes (point mutations, deletions) that alter the properties of the protein chains which make up the hemoglobin tetramer, or alter the proportion of the constituent chains. These alterations change the electrophoretic mobilities of the variants such that they can be resolved by capillary electrophoresis. Capillary zone electrophoresis has been applied to separation of some hemoglobin variants; for example, Klein and Jollif [270] resolved hemoglobins A, A_1, A_2, S, and C by CZE using a barbital buffer of undisclosed composition. However, because of the large number of known hemoglobin variants (over 500), glycated hemoglobins and normal hemoglobin species, CZE does not have sufficient resolving power to serve as a routine method for differentiating Hb variants. The subtle structural changes in Hb variants often causes slight changes in their isoelectric points which can be readily distinguished by isoelectric focusing, and CIEF appears to be a more promising approach for Hb variant analysis (see Chapter 6). On the other hand, point mutations can alter the electrophoretic mobility of the individual globin chains such that aberrant forms can be resolved from normal chains. For example, Ferranti et al. [271] employed CZE under acidic conditions (20 mM phosphate, pH 2.5) with coated capillaries to resolve globin chains carrying acidic-to-neutral (Hb S, glu > val; Hb San Jose, glu > gly) or acidic-to-basic (Hb C, glu > lys; Hb O-Arab, glu > lys) amino acid substitutions. Ong et al. [272] used CZE under basic conditions (25 mM phosphate, pH 11.8) with uncoated capillaries to identify altered α/β chain ratios in β-thalassemia carriers. Capillary zone electrophoresis under denaturing conditions was used by Zhu et al. [273,274] to identify altered globin chains in hemoglobin variants (Hb S, C, E, and G Philadelphia) and altered globin chain ratios in β-thalassemia samples (Hb H, Hb Bart's). In these studies, a 100 mM sodium phosphate at pH 3.2 containing 7 M urea + 1% reduced Triton X-100 was used (Figure 24).

Lipoproteins

Serum lipoproteins have been analyzed by isotachophoresis, typically after staining with a lipophilic dye such as Sudan Black [275,276]. Apolipoproteins can be analyzed by CZE as reported by Tadey and Purdy [277,278]. They were able to resolve apoA-I, apoA-II, apoB-100, and apo-B48 from HDL and LDL preparations using uncoated capillaries with a 30 mM borax buffer (pH 9) containing SDS. Other detergents were less effective, although either SDS or cetyl trimethylammonium bromide provided good resolution of VLDL apolipoproteins with polyacrylamide-coated capillaries. Lehmann et al. [261] developed a method for direct analysis of apoA-I in serum using an uncoated 50 cm × 50 µm capillary and a proprietary buffer. Dilution of the serum sample in the buffer allowed direct injection and resolution of apoA-I from all other serum proteins, and apoA-I levels in patient sera correlated well with values from nephelometric determinations. Using a 50 mM borate buffer containing 3.5 mM SDS and 20% (v/v) acetonitrile, Cruzado et al. [279,280] compared CZE with reversed phase HPLC for separation of apolipoproteins A-I and A-II. The HPLC method could resolve apoA-I and apoA-II into 3 and 2 isoforms, respectively, while CZE could not. However, the apoA-I and apoA-II isoforms overlapped and could not be resolved by HPLC, preventing analysis of the two apolipoproteins in mixtures. Therefore, CZE was the preferred technique for quantitation of apoA-I and apoA-II in HDL. Capillary electrophoresis values for these two proteins determined in nondelipidated HDL fractions obtained from serum controls by density gradient centrifugation agreed well with immuno-based assay values.

Urinary Proteins

Jenkins et al. [263] analyzed proteins in urine using a 150 mM borate buffer (pH 9.7) containing 1 mM calcium lactate; the higher borate concentration was used to supress protein-wall interactions.

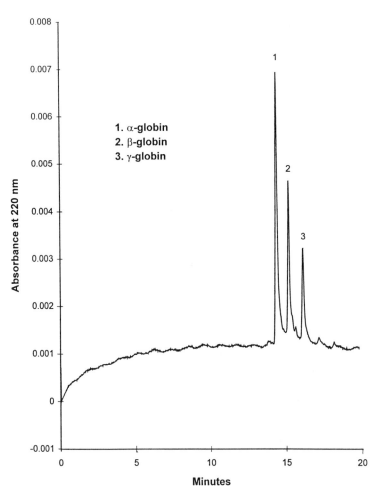

Figure 24 Separation of globin chains under denaturing conditions. Globins were precipipated from blood with acidic acetone (2% concentrated HCl in acetone), washed twice with acetone and reconstituted in 10 mM sodium phosphate buffer (pH 3.2) + 7 M urea + 0.1% reduced Triton X-100. Analysis conditions: capillary, 35 cm × 25 μm coated with LPA; injection, 8 kV for 8 s; analysis buffer, 100 mM sodium phosphate (pH 3.2) + 7 M urea + 1% reduced Triton X-100; polarity, positive to negative; voltage, 8 kV.

Urine samples were pretreated using ultrafiltration concentrators with a 15 kDa cut-off to achieve a 100-fold increase in protein concentration and to eliminate low-molecular-weight components. Quantitative comparison of CZE with HRAGE yielded a correlation coefficient of 0.93 for albumin and 0.95 for Bence Jones protein.

Use of CZE to analyze myoglobin in urine for detection of myoglobinuria was reported by Shihabi [281]. Using a 150 mM borate buffer (pH 8.7) containing 0.5% PEG and detection at 405 nm, myoglobin was resolved into two species, MI and MII. The MII peak was converted to MI after incubation of samples at room temperature. Using hemoglobin as an internal standard, myoglobin values determined by this method agree with anion exchange HPLC values.

6
Capillary Isoelectric Focusing

INTRODUCTION
Equilibrium Techniques

The natural end of an analysis is when resolution of the sample components of interest has been achieved. In practice, however, most assays are terminated arbitrarily. The end point of the experiment is determined by the researcher, and it is usually based on factors that include time and space (e.g., before the proteins of interest exit the gel in slab gel electrophoresis, or after all analytes have eluted from a chromatographic column). But there is also a small group of techniques in which the end point is determined by the attainment of equilibrium. These techniques are collectively called isoperichoric, meaning "equal to its environment" [282].

The main common characteristic of equilibrium techniques is that the analysis proceeds until equilibrium is reached, and no further resolution is achieved after this stage has been accomplished. This fact is of prime relevance to the understanding, optimization, and troubleshooting of these methods. Some of the most popular isoperichoric methodologies include density gradient centrifugation or equilibrium centrifugation, and isoelectric focusing. All these techniques rely upon the generation of a specific gradient in which sample molecules can be separated. The molecular differences exploited by these methods include density (centrifugation) and isoelectric point (isoelectric focusing). An interesting feature of equilibrium methods is that usually separation and concentration occur at the same time, which is a departure from most other techniques in which dilution occurs during the analysis.

Isoelectric focusing is applicable only to molecules which possess isoelectric points. However, it is difficult to analyze small zwitterions in conventional gel IEF due to the lack of suitable detection methods and the high diffusion coefficient of the analytes. Consequently, isoelectric focusing has been applied almost exclusively to the analysis of proteins and peptides.

Although analysis of amphoteric molecules can be traced back to 1912 when Ikeda and Suzuki used a three-compartment apparatus to produce glutamate, and to 1929 when Williams and Waterman developed a remarkable separation concept based on pI using a 12-compartment apparatus, the basis for modern IEF was established first by Kolin with his artificial pH gradients, and then by Svensson and Vesterberg with the introduction of the concept and material production of carrier ampholytes [282], which generated natural gradients. Along the way, many other concepts were introduced, but the key was to increase the number of compartments as a means to improve the power of resolution, i.e., to generate a smoother pH gradient. Perhaps Kolin's main contribution [283] was that the pH gradient could be obtained in a single chamber by diffusion of two or more buffers ("artificial" pH gradients) coupled with a density gradient. Obviously, these gradients were short lived, and thus Svensson theorized that a series of amphoteric buffers driven to their pI by the electric current would form a more stable pH gradient ("natural" gradients). When Vesterberg synthesized carrier ampholytes, all the main elements were in place for the development of IEF as a powerful analytical tool.

Capillary IEF

Isoelectric focusing is widely used for analysis and characterization of polypeptides in slab gel matrices that eliminate convection and provide a support for sample staining. The importance of IEF as an analytical tool is not only that it provides a powerful means to determine sample composition, but also that it can be used to measure the isoelectric point of complex analytes.

With the development of capillary electrophoresis (CE) as an alternative to slab gels to eliminate convection problems, IEF was successfully transferred to the capillary format. Capillary isoelectric focusing (CIEF) combines the high resolving power of conventional gel isoelectric focusing with the advantages of capillary electrophoresis instrumentation. Just as in gel IEF, proteins are separated according to their isoelectric points in a pH gradient generally formed by carrier *ampholytes* (*ampho*teric electro*lytes*) when an electric potential is applied. The use of small diameter capillaries allows the efficient dissipation of Joule heat and permits the application of high voltage for a rapid focusing of the protein zones. In CE, separations are commonly performed in free solution, i.e., in the absence of any support such as gel matrices. This allows the replacement of the capillary's content in between analyses and, therefore, the automation of the process. The use of UV-transparent fused-silica capillaries enables direct on-line optical detection of focused protein zones, eliminating the requirement for sample staining. The resolving power of CIEF is usually higher than most protein analysis techniques, including other modes of capillary electrophoresis. The introduction of CIEF expanded the use of IEF to include the analysis of peptides, amino acids, and other small organic zwitterions.

As in conventional IEF, the high resolving power of CIEF depends upon the focusing effect of the technique (Figure 1). At steady state, the ampholytes form a stable pH gradient within which proteins become focused at the position where their net charge is zero, i.e., where pH = pI. Diffusion of a protein toward the anode (positive electrode) will result on the acquisition of positive charge, resulting in the molecule returning to the focused zone (attracted by the cathode). Similarly, diffusion toward the cathode (negative electrode) will result in acquisition of negative charge, causing back-migration to the pI zone. As long as the electric field is applied, electrophoretic migration counters the effects of diffusion. Since detection in CIEF is performed on-line, the electric field is maintained throughout the analysis, and resolution is usually very high. Capillary IEF usually produces more complex patterns than conventional IEF because smaller peptides (and even some amino acids) that do not stain well or diffuse out of the

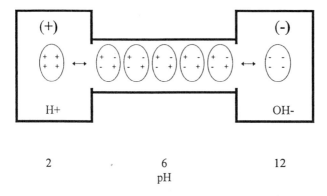

Figure 1 Focusing phenomena during isoelectric focusing. Amphoteric molecules placed between a low pH anolyte and a high pH catholyte acquire a net charge when diffusing to either of the flanking solutions. In the presence of an electric field they migrate back to their isoelectric point. If the amphoteric solution has buffering capacity, it maintains the pH at that particular value in the area it occupies.

gel are also detected. Capillary IEF offers resolution comparable to slab gel IEF, but not as high as immobilized pH gradients [284].

For all CE instruments that use on-line detection at a fixed point along the capillary, CIEF must include a means of transporting the focused zones past the detection point. This process, commonly referred to as mobilization, can occur as an independent stage or be combined with the focusing of the sample-ampholyte matrix components. The forces used to achieve mobilization are various, and in some instances they are applied in combinations of two or more simultaneously.

There are several protocols for performing capillary isoelectric focusing, but generally they can be classified in two groups, two-step CIEF and single-step CIEF. Two-step CIEF is characterized by the performance of focusing and mobilization as two distinct stages. Mobilization can be achieved by ion addition, or by applying a hydraulic force (pressure, vacuum, gravity). This method requires electroosmosis to be eliminated (or reduced to a very low level). In single-step CIEF, focusing occurs while the nascent protein zones are being transported toward the detection point. The forces used to transport the focused zones are the same as those for multiple-step CIEF, but in this case electroosmotic flow (EOF) can also be used, alone or in combination with any of the other forces.

Each of these approaches requires specific instrument configurations and different strategies for optimization of the CIEF process, and they will be discussed separately. Since sample injection and focusing are common to all variations of CIEF, these stages will be described first.

SAMPLE PREPARATION AND INJECTION

Sample preparation for CIEF includes selection of the appropriate ampholyte composition, adjustment of sample salt levels, and dilution or concentration of the sample to the proper protein levels required for detection.

Sample Salt Content

The ionic strength of the sample should be as low as possible, with a practical higher limit around 30–50 mM. Excessive sample ionic strength due to the presence of salts (including buffer) or ionic detergents will interfere with the isoelectric focusing process, greatly increasing focusing times and causing peak broadening during mobilization. Elevated current due to the presence of salt can increase the risk of precipitation as proteins become concentrated in focused zones. Samples with salt concentration of 50 mM or greater should be desalted by dilution, dialysis, gel filtration, or ultrafiltration.

Ampholyte Composition

The ampholyte composition should be selected based upon the desired separation range. For separating complex samples containing proteins with widely different isoelectric points, or to estimate the pI of an unknown protein, a wide-range ampholyte blend should be selected, e.g., pH 3–10. The final ampholyte concentration should be between 1 and 2%. In order to detect proteins at the basic end of the gradient during cathodic mobilization, it is necessary that the pH gradient span only the effective length of the capillary, e.g., the distance from the capillary inlet to the detection point. In cases where the total capillary length is much greater than the effective length, many sample components may focus in the "blind" segment distal to the monitor point and be undetected during mobilization. A basic compound such as N,N,N',N'-tetramethylethylenediamine (TEMED) can be used to block the distal section of the capillary (Figure 2). As a rule of thumb, the ratio of TEMED concentration (% v/v) to ampholyte concentration should be approximately equal to the ratio of the "noneffective" (or "blind") capillary length to total length. For example, if the effective length of a 25 cm capillary is 20 cm, a 0.4% concentration of TEMED should be added if the final ampholyte concentration is 2%. Note that the use of TEMED is required only if the sample contains proteins focusing at the basic end of the gradient in cathodic mobilization (e.g., proteins with pIs above 8–8.5 when using a pH 3–10 ampholytes). Use of TEMED for CIEF of proteins with pI values below 8.5 will have no beneficial effect and, in fact, can reduce resolution by increasing the slope of the pH gradient. Limiting the range of the pH gradient can also improve the lifetime

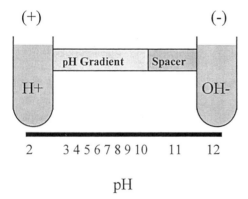

Figure 2   The use of spacers in CIEF. A basic spacer can be used to block the distal section of the capillary allowing detection of proteins that otherwise would focus in this "blind" segment of the column. Ideally, the spacer occupies the segment between the detection point and the outlet of the capillary.

of coated capillaries, and thus the overall performance of the CIEF process.

In situations where enhanced resolution of proteins with similar pI values is desired, the use of narrow-range ampholyte mixtures may be considered. Narrow-range ampholyte mixtures generating gradients spanning 1–3 pH units are available from several commercial sources. However, our experience with this approach to high-resolution CIEF has been disappointing, perhaps due to the limited number of ampholyte species in narrow-range "cuts." We have obtained better results by adding various ratios (10–80%) of narrow-range ampholytes to a "base" of broad-range ampholytes. Since the pH gradient, and thus, the resolving power of IEF depends on the number of ampholyte species present, one proposed solution to increase resolution has been to blend narrow-range ampholytes from several manufacturers [285], and also from several production batches.

Protein Concentration

The final protein concentration in the sample + ampholyte mixture will depend on sensitivity requirements and the solubility of protein components under focusing conditions. As an approximation, a final concentration of 0.5 mg/ml per protein should provide adequate sensitivity and satisfactory focusing + mobilization performance. However, many proteins may precipitate during focusing at this starting concentration, since the final protein concentration in the focused zone may be as high as 200 mg/ml. Immunoglobulins, membrane proteins, and high-molecular-weight or hydrophobic proteins, in general, have a higher risk of precipitation in CIEF. In such cases, the use of very dilute protein solutions may be required. Prior to injection, the prepared sample should be centrifuged for 2–3 minutes in a microcentrifuge to remove any particulate material and to degas the solution. This practice is particularly important if the sample may contain protein aggregates or other large particles, and when polymers are used to increase the viscosity of sample + ampholytes solution.

Once prepared, the sample + ampholyte mixture is introduced into the capillary. This can be accomplished by pressure injection or by vacuum. For good quantitative precision, sufficient volume should be loaded into the capillary to ensure that the tube contains a homogeneous mixture of sample. Therefore, the capillary should be injected with at least 3–5 volumes of sample + ampholytes (for two-step-CIEF). In contrast to other modes of CE, the injection volume in CIEF usually occupies from several centimeters to the whole length of the capillary.

FOCUSING

There are several approaches that can be used to create a pH gradient. However, we will describe only techniques readily available to the everyday CE practitioner. To date, the most common approach to the generation of a pH gradient is through the use of carrier ampholytes.

The Role of Ampholytes

Ampholytes are mixtures of a high number of synthetic chemical species that posses slightly different pIs. Carrier ampholytes are oligoamino-oligocarboxylic acids with different pI values [286]. These compounds were first obtained by Vesterberg by randomly polymerizing pentaethylenehexamine and acrylic acid, thus generating thousands of ampholytes with different pI values. The number of ampholytes per pH unit has been calculated to range between 50 and 1000. It has also been determined that in order to have a smooth pH slope, there should be at least 30 ampholytes per pH unit, otherwise a step pH gradient is obtained. Besides having different pI values, carrier ampholytes must be good buffers and also good conductors at their pI, so that they can carry the electric current and at the same time maintain the pH gradient steady.

Explaining the role of ampholytes can be greatly simplified if we consider the following hypothetical situation (Figure 3A). If a capillary is filled with a solution containing an amphoteric substance (e.g., ampholyte, protein, etc.) that is flanked by an acidic solution (pH below the ampholyte's pI) at the anode and an alkaline solution (pH

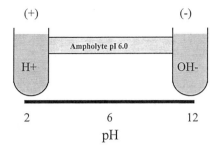

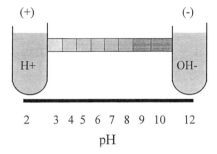

Figure 3 Creation of a pH gradient by ampholytes. Top: When the capillary is filled with a single ampholyte, the solution maintains the pH at a value dependent on the ampholyte used (in this example the pH is 6.0). Bottom: A mixture of ampholytes produces a gradient (if the number of ampholytes is low the result is a step gradient; if the number of ampholytes is high the gradient is smoother). To achieve separation, at least one ampholyte must possess a pI value intermediary to the two peaks to be resolved.

above the ampholyte's pI) at the cathode, upon applying an electric field, all molecules with a net charge migrate toward the electrode of opposite charge. However, because the pH outside the capillary is greater (or lower) than their pI, the molecules will not exit the capillary (due to loss of charge or charge reversal). If the ampholyte possesses buffer capacity, it will maintain the pH inside the capillary to a determined value (pH 6.0 in Figure 3A). A similar situation results if a mixture of ampholytes with different pIs is used, and the net result is the creation of a pH gradient (Figure 3B).

The Focusing Process

During the performance of a CIEF analysis, the focusing step begins with the immersion of the capillary in the anolyte (dilute phosphoric acid) and catholyte (dilute sodium hydroxide) solutions, followed by application of high voltage. Typically the catholyte solution is 20–40 mM NaOH, and the anolyte is half the catholyte molarity, e.g., 10–20 mM phosphoric acid. It is important that the catholyte be prepared fresh, as sodium hydroxide solutions will gradually take up carbon dioxide from the atmosphere. The presence of carbonate salts in the catholyte will interfere with the focusing process. Use of higher NaOH concentrations (e.g., 40 mM) minimizes this problem. However, the concentration of NaOH should not be so high that it would start to dissolve the silica that makes up the capillary. For narrow bore capillaries (e.g., 50 μm i.d.) , field strengths of 300-900 V/cm or greater can be used. Our experience indicates that about 600 V/cm is optimal. Upon application of high voltage, the charged ampholytes migrate in the electric field. A pH gradient begins to develop, with low pH toward the anode (+) and high pH toward the cathode (−); the range of the pH gradient in the capillary is defined by the composition of the ampholyte mixture. At the same time, protein components in the sample migrate until a steady state is reached, at which point each protein becomes focused in a narrow zone at its isoelectric point. Focusing is achieved rapidly (typically within a few minutes in short capillaries) and is accompanied by an exponential drop in current. Current drop can be explained as follows: when the electric field is applied at the beginning of focusing, most components in the sample matrix have net charge and act as current carriers. When ampholytes and polypeptides reach their isoelectric point, their net charge changes to zero and, therefore, the value of the current declines. Focusing is usually considered to be complete when the current has dropped to a level approximately 10% of its initial value (for samples containing low salt) and the rate of change approaches zero. It is generally not advisable to prolong focusing beyond this point, since the risk of protein precipitation increases with time. Also, loss of ampholytes at the acidic or basic ends of the gradient can give rise to anodic or cathodic drift [287]. Anodic drift can be minimized by increasing the phosphoric acid concentration of the anolyte [288].

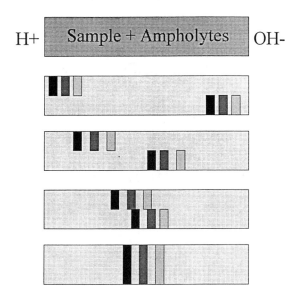

H+ Sample + Ampholytes OH-

Figure 4 Focusing process during CIEF. The capillary is filled with a protein–ampholyte mixture. The ends of the capillary are then immersed in anolyte (H^+) and catholyte (^-OH). Upon the application of an electric field the ampholytes and protein migrate toward their isoelectric point. Protein zones are formed at both ends of the gradient. The protein zones at both ends of the gradient converge at the point where the pH equals their pI. For clarity, only one half of each band formed at the capillary ends is depicted in the middle three panels.

The focusing process can be monitored by the movement of nascent protein zones past the detection point (Figure 4). The detection profile generated during this step may not exhibit well-resolved peaks, but it is reproducible and characteristic of the sample. In cases where protein precipitation prevents acquisition of reliable data during mobilization, the focusing profile can sometimes yield useful information about the sample [289]. Nascent zones are generated at both ends of the protein–ampholyte mixture, and eventually they merge at a point in the capillary where they are isoelectric. If the whole capillary has been filled with sample and ampholytes, a drop in the baseline can be noticed after the focusing protein zones pass the detector point. This occurs because when the analysis is initiated, the instrument automatically "zeroes" the detector baseline while proteins and ampholytes are in the light path; when these compounds move to some point in the pH gradient, more light reaches the photodiode and, therefore, the baseline "sinks." As a rule of thumb, for protocols that achieve complete focusing, whenever the length of the applied sample zone exceeds the effective length of the capillary, focusing peaks should be detected during the analysis.

If focusing time or field strength is not sufficient, coalescence of the nascent zones may be incomplete, causing a single protein species to appear as an unresolved doublet or as separate peaks. These may be mistakenly identified as separeate proteins in the mobilization stage (Figure 5).

pH Gradient Instability

Throughout this description of the IEF process we have referred to IEF as a technique in which equilibrium is attained when the pH gradient is formed. In practice, however, this steady state is compromised by several phenomena [286,290]. One well-studied effect is the *plateau phenomenon*, a process by which the gradient flattens out in the neutral region and becomes steeper in the acidic and basic regions. This phenomena is a consequence of electroneutrality. For example, consider a hypothetical ampholyte with a pI = 10.0. The hydroxide ion concentration in the zone of this ampholyte is 100 μM. Thus, owing to electroneutrality, there must be an excess concentration of the positively charged form of the ampholyte over the negatively charged one of the same amount (in this case 100 μM). These electrically charged ampholytes migrate under the influence of the applied electric field, exiting the capillary despite the high pH of the catholyte solution (in which direction the ampholyte migrates). Acidic ampholytes encounter a similar situation, with the charge and the migration direction reversed (with the excess negatively charged ampholyte migrating toward the anode). Only ampholytes in the neutral region will not have a significant amount of one charged form over the other.

One thing that we have to keep in mind is that ampholytes focused in a pH zone at their pI exhibit no net charge, but there are different

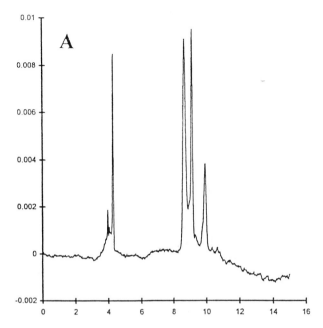

Figure 5 Multiple peaks due to incomplete focusing. Horse heart myo-globin was diluted to 0.1 mg/ml in Pharmalyte 3-10 (5% v/v) + 0.3% meth-ylcellulose (4K cps at 2%). Focusing was carried out using a voltage ramp from 0 to 15 kV/5 min. Hydraulic mobilization (0.5 psi) was started at different time intervals after the completion of the voltage ramp. (A) Zero time after completion of voltage ramp. (B) 1 min at 15 kV after completion of voltage ramp. (C) 2 min at 15 kV after completion of voltage ramp. Capillary: 25 cm × 50 μm, polyAAEE-coated (Bio-Rad Laboratories), thermostated to 25°C. Detection at 415 nm.

charged forms. Since ampholytes at both ends of the pH gradient exit the capillary, the neutral region slowly expands. The net effect of this process is that the resolution of the polypeptides in the neutral region of the pH gradient increases with time, and the resolution at the extremes of the pH decreases.

Two other related causes of pH gradient instability are known as *cathodic and anodic drift*. This phenomena also refers to the loss of basic ampholytes into the cathodic reservoir, and acidic ampholytes

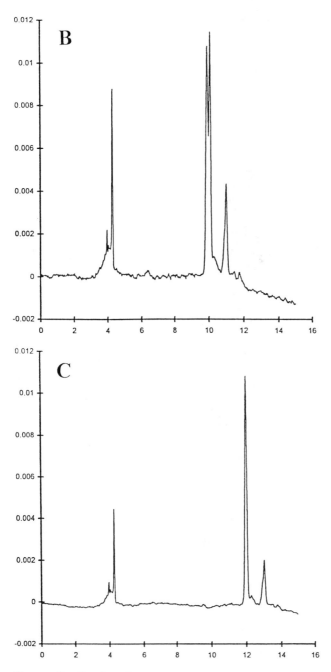

Figure 5 Continued

into the anodic reservoir, as already discussed for the plateau phenomena. Thus, these drifts are isotachophoretic phenomena. Isotachophoresis (ITP) is an electrophoretic phenomena based on ionic species assuming the same velocity and concentration of a leading ion of high mobility. In anodic drift, the anion in the anolyte acts as an isotachophoretic leader for the negatively charged ampholytes in the acidic region, whereas the cation in the catholyte does the same for the cationic ampholytes in the basic region. When equal concentrations of phosphoric acid and sodium hydroxide (i.e., the concentration of the providers of leading ions) are used as anolyte and catholyte, the drift occurs mainly toward the cathode. This effect is due to the higher mobility of the sodium as compared to the phosphate ion. If the concentration of the sodium hydroxide is increased to 2.25 times that of the phosphoric acid, symmetrical drifts occur. If the concentration of NaOH exceeds 2.25 times the phosphoric acid concentration, the drift will be mainly anodic. In gels, it has been shown that cathodic drift can be minimized by choosing an anolyte with a pH similar to the pI of the most acidic ampholyte [282]. For an ampholyte range of 6–8, this was achieved by using threonine as anolyte and histidine as catholyte.

TWO-STEP CIEF

Since commercial CE instruments use on-line detection at a fixed point along the capillary, CIEF must include a means of transporting the focused zones past the detection point. Mobilization has been regarded for the most part as a stage of little importance in the overall performance of the CIEF process, but now it has been shown that mobilization conditions can be manipulated to improve resolution and reproducibility. Three approaches have been used to mobilize focused zones. In *chemical mobilization* (ion addition), changing the chemical composition of the anolyte or catholyte causes a shift in the pH gradient, resulting in electrophoretic migration of focused zones past the detection point [291,292]. In *hydraulic mobilization*, focused zones are transported past the detection point by applying pressure [293,294], or vacuum [288] at one end of the capillary, or by volume height differential of the anolyte and catholyte levels (siphon or gravity) [295]. In *electroosmotic mobilization*, focused zones are transported past the detection point by electroosmotic pumping [287,296,297]. Mobilization by EOF is used in single step CIEF only.

Chemical Mobilization

At the completion of the focusing step, high voltage is turned off and the anolyte or catholyte is replaced by the mobilization reagent (Figure 6). High voltage is again applied to begin mobilization. As in focusing, field strengths of 300–900 V/cm can be used for mobilization, with optimum separations achieved in small i.d. capillaries using a field strength of about 600 V/cm. The choice of anodic vs. cathodic mobilization and the composition of the mobilizing reagent depends upon the isoelectric points of the protein analytes, and the goals of separation. Since the majority of proteins have isoelectric points between 5 and 9, cathodic mobilization (mobilization toward the cathode) is most often used.

Mobilization Using NaCl

To date, the most common chemical mobilization method is the addition of a neutral salt such as sodium chloride to the anolyte or catholyte; sodium serves as the nonproton cation in anodic mobilization and chloride functions as the nonhydroxyl anion in cathodic mobilization. A suggested cathodic mobilization reagent is 80 mM NaCl in 40 mM NaOH. At the beginning of mobilization, current initially remains at the low value observed at the termination of focusing, but gradually begins to rise as the chloride ions enter the capillary. Later in mobilization, when chloride is present throughout the tube, a rapid rise in current signals the completion of mobilization (Figure 7A). The electrical current at the end of mobilization using NaCl is much higher than the current observed at the beginning of focusing.

Ideally, mobilization should cause focused zones to maintain their relative position during migration, i.e., zones should be mobilized as a train past the monitor point. In practice, movement of ions into the capillary causes a pH change at the capillary end which progresses deeper into the tube. The rate of change depends upon the amount of co-ion moving into the capillary, the mobility of the co-ion, and the

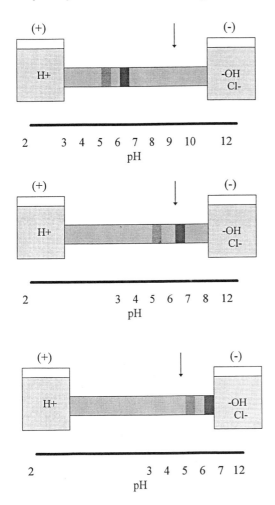

Figure 6 Schematic of mobilization by ion addition. A competing ion is added (Cl⁻) to one of the focusing reagents (in this case the catholyte) disrupting the equilibrium attained during focusing. The addition of the competing ion causes a pH shift that sweeps the entire length of the capillary, effectively mobilizing the focused protein zones.

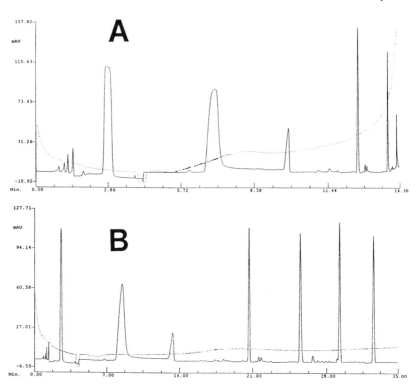

Figure 7 Monitoring current during ion-addition mobilization. The solid
line represents detection of BioMark pI markers (Bio-Rad Laboratories);
the dotted line traces the current behavior during the analysis. The initial
current depends on concentration of salts present in the sample solution,
the concentration of ampholytes, and the concentration of sample components. During focusing the current drops rapidly (first 4 min in both panels). (A) Ion-addition mobilization with NaCl. The current rises slowly at
the beginning, and then sharply toward the end of mobilization. In this
particular analysis, the current increased dramatically (>100 μA) before
all the sample components were detected. (B) Ion-addition mobilization
with zwitterion solution. Note that the final current does not increase as
much as with NaCl mobilization. Peaks: (1) BioMark 10.4, (2) BioMark
8.4, (3) BioMark 7.4, (4) BioMark 6.4, (5) BioMark 5.3.

buffering capacity of the carrier ampholytes. The actual slope of the pH gradient changes across the capillary, becoming shallower in the direction opposite to mobilization (Figure 8). Neutral and basic proteins are efficiently mobilized toward the cathode with sodium chloride, and mobilization times correlate well with pI (Figure 9). However, acidic proteins at the far end of the capillary are mobilized with lower efficiency and may exhibit zone broadening or be undetected.

Mobilization Using Zwitterions

Use of zwitterions is an alternative approach which provides more effective mobilization of protein zones across a wide pH gradient [298]. For example, cathodic mobilization with a low-pI zwitterion enables efficient mobilization of proteins with pIs ranging from 4.65 to 9.60. The proposed mechanism for zwitterion mobilization couples a pH shift at the proximal end of the tube with a displacement effect at the distal end as the zwitterion forms an expanding zone within the gradient at its isoelectric point. Effective zwitterion mobilization depends on selection of the appropriate mobilization reagent. For example, cathodic mobilization requires a zwitterion with an isoelectric point between the pH of the anolyte and the pI of the most acidic protein. The current level increment during zwitterion mobilization is less than that observed in salt mobilization (Figure 7B). Selective mobilization can be accomplished by using a zwitterion with a pI value just lower than the compound of interest. In this case, only proteins with a higher pI than the used ampholyte are mobilized.

Principle of Chemical Mobilization

The theoretical basis for chemical mobilization was described by Hjertén et al. [292]. At steady state, the electroneutrality condition in the capillary during focusing can be expressed as

$$C_{H^+} + \Sigma C_{NH_3^+} = C_{OH^-} + \Sigma C_{COO^-} \tag{6-1}$$

where C_{H^+}, C_{OH^-}, $C_{NH_3^+}$, and C_{COO^-} are the concentrations of protons, hydroxyl ions, and positive and negative groups in the ampholytes, respectively. In anodic mobilization, addition of a nonproton cation X^{n^+} to the anolyte introduces another term to the left side of the equation:

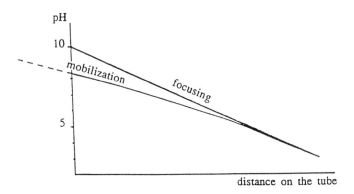

Figure 8 Slope of the pH gradient during CIEF. Notice that during focusing the gradient is linear along the axis of the capillary, but during ion-addition mobilization using NaCl the "modified" gradient is shallow in the side opposite to mobilization (Reproduced from Ref. 298 with permission.)

Figure 9 Correlation of isoelectric point and migration time for the various types of mobilization. (A) Calibration curve for CIEF with chemical mobilization achieved by adding 30 mM NaCl to the catholyte (20 mM NaOH). Focusing was performed in a 27 cm × 50 μm μ-SIL DB-1 capillary (J&W Scientific) at 20 kV for 5 min. Phosphoric acid at 20 mM was used as anolyte. The carrier ampholyte solution consisted of 4% Pharmalyte 3-10, 1% TEMED and 0.8% methylcellulose. (B) Linearity plot for pressure mobilization. Analysis conditions identical to those of A except that mobilization was performed by applying 0.5 psi while maintaining the voltage at 10 kV. (C) EOFmobilization. Analysis was performed in an eCAP neutral capillary (Beckman Instruments) at 10 kV using reverse polarity at 10 kV. Anolyte was 10 mM phosphoric acid and catholyte was 20 mM NaOH. The carrier amophoyte solution contained 4% Pharmalyte 3-10, 1.5% TEMED, and 0.4% methycellulose. (Reproduced from Ref. 302 with permission.)

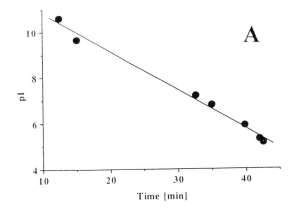

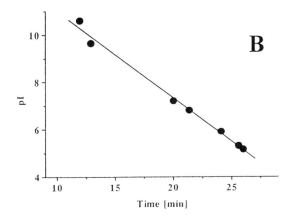

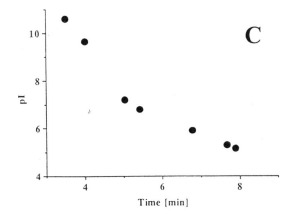

$$C_{X_n^+} + C_{H^+} + \Sigma C_{NH_3^+} = C_{OH^-} + \Sigma C_{COO^-} \tag{6-2}$$

Migration of the nonproton cation into the capillary will result in a reduction in proton concentration, i.e., an increase in pH. Similarly, addition of a nonhydroxyl anion Y^{m-} to the catholyte in cathodic mobilization yields a similar expression:

$$C_{H^+} + \Sigma C_{NH_3^+} = C_{OH^-} + \Sigma C_{COO^-} + C_{Y_m^-} \tag{6-3}$$

indicating that migration of a nonhydroxyl anion into the capillary results in a reduction in hydroxyl concentration, i.e., a decrease in pH. Progressive flow of nonproton cations (anodic mobilization) or nonhydroxyl anions (cathodic mobilization) will, therefore, cause a progressive pH shift down the capillary, resulting in mobilization of proteins in sequence past the detector point. The pH shift occurs first near the end of the capillary, but because small ions, like ^-OH, H^+, Na^+, Cl^-, etc., have a very high mobility, the pH transition is propagated quickly to the rest of the gradient. If snapshots could be used, the pH changes would be seen as pulses that sweep the capillary from one end to the other, altering the pH gradient as they move along. In reality, the changes are not pulses, but a continuous process. Figure 6 represents "snapshots" at different times during mobilization.

Hydraulic Mobilization

Hydraulic mobilization utilizes positive pressure or negative pressure (vacuum) as the force that transports the focused protein zones toward the detection point. During hydraulic mobilization, it is necessary to apply an electric field across the capillary in order to maintain focused protein zones [288]. The main disadvantage of this type of mobilization is the parabolic shape of the hydrodynamic flow profile, which can decrease resolution. For this reason, only weak forces are used.

Pressure Mobilization

Capillary isoelectric focusing with hydraulic mobilization was first described by Hjertén and Zhu [293]. Mobilization was accomplished by displacing focused zones from the capillary by pumping anolyte solution into the capillary using an HPLC pump equipped with a T-connection to deliver a flow rate into the capillary of 0.05 μl/min. Voltage was maintained during mobilization to keep the protein zones focused, and on-tube detection using a UV detector was employed. Since their pioneering studies, other forms of pressure have been used, e.g., compressed gas [294], pressure created by height difference of liquid levels contained in the reservoirs (we refer to this as "gravity mobilization"), and vacuum [288].

Gravity Mobilization

From an instrument perspective, the simplest hydraulic approach to transport focused zones to the detector is by gravity mobilization [295]. In this technique, after focusing the proteins are transported toward the detection point using a difference in the levels of anolyte and catholyte contained in the reservoirs (Figures 10 and 11). The force generated by the liquid height difference can be manipulated to be extremely small compared with pressure or vacuum, and flow velocity can be modulated by changing the capillary dimensions or (in the case of large-bore capillaries with internal diameters greater than 50 μm) by addition of viscous polymers.

Vacuum Mobilization

Hydraulic mobilization by vacuum and using on-line detection was described by Chen and Wiktorowicz [288]. In this approach, a four-step vacuum-loading procedure was used sequentially to introduce segments of catholyte (20 mM NaOH + 0.4% methylcellulose), ampholytes + methylcellulose, sample solution, and a final segment of ampholytes + methylcellulose from the anodic end of the capillary. Following loading of the column, focusing was carried out for 6 min at a field strength of 400 V/cm; then mobilization of focused zones toward the cathode was performed by applying vacuum at the capillary outlet with voltage simultaneously maintained to counteract the distorting effects of laminar flow. A dimethylpolysiloxane (DB-1) coated capillary (J & W Scientific, Folsom, CA) combined with addition of methylcellulose to the catholyte and ampholyte solutions served to suppress EOF. Relative mobility values for proteins were calculated by normalizing zone migration times to the migration times

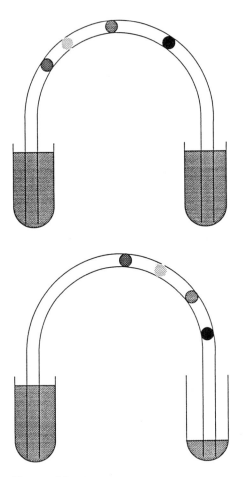

Figure 10 Schematic of gravity mobilization. Once equilibrium has been attained during focusing, transport of protein zones toward the detection point is accomplished by applying a height difference between anolyte and catholyte levels. To minimize the loss of resolution, the voltage must be maintained during mobilization.

of the catholyte–ampholyte and ampholyte–anolyte interfaces. A plot of relative mobility values vs. pI values of protein standards was linear over a pH range of 2.75–9.5. Using this approach, nuclease isoforms were successfully analyzed by CIEF [299] which could not resolved satisfactorily by other modes of CE (CZE and emulsion MECC).

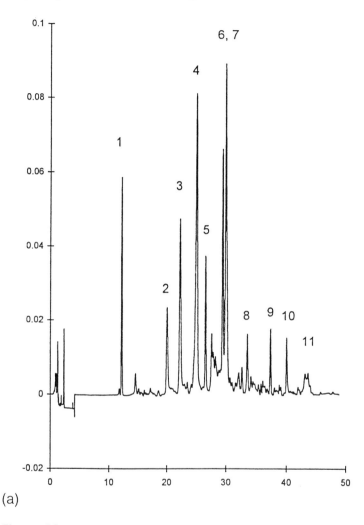

(a)

Figure 11 Capillary IEF of standard proteins using (a) gravity mobilization and (b) ion-addition mobilization. The sample was diluted 49:1 with
2% ampholytes 3–10 containing 0.5% TEMED to block the blind side of
the capillary. Proteins were focused for 4 min at 15 kV using a 30cm × 75
μm coated capillary. Gravity mobilization was implemented using a height
difference of about 2 cm while maintaining the high voltage on (15 kV) to
prevent sample diffusion and defocusing. Ion-addition mobilization was
accomplished using a zwitterion solution. The capillary and all solutions
were thermostatted at 20°C. Peak identities: (1) Cytochrome C (2, 3, 4)
lentil lectins; (5) human hemoglobin C; (6) human hemoglobin A; (continued)

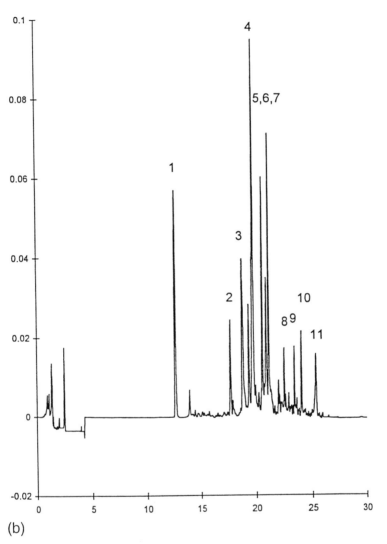

(b)

Figure 11 (Continued) (7) horse heart myoglobin; (8) human carbonic anhydrase; (9) bovine carbonic anhydrase; (10) lactoglobulin B; (11) phycocyanin.

SINGLE-STEP CIEF

Single-step or dynamic CIEF is a variation of CIEF in which focusing and transport of the sample occurs simultaneously. Single-step CIEF was first developed as a means to perform CIEF in uncoated capillaries. Later, dynamic CIEF using pressure (gravity) was demonstrated. The advantage of single-step CIEF is the simplification of the protein pattern (there are no focusing peaks), but because the capillary is only partially filled with sample-ampholytes solution some resolution and sensitivity are sacrificed.

Single-Step CIEF in the Presence of EOF

Isoelectric focusing using capillaries with significant levels of EOF is a one-step process, with focusing occurring while sample proteins are being transported toward the detection point by electroosmotic flow. This technique has been used with both uncoated capillaries and with capillaries coated to reduce (but not eliminate) EOF.

Two approaches have been reported for single-step CIEF in the presence of EOF, one in which the sample + ampholyte mixture was introduced as a plug at the inlet of the capillary prefilled with catholyte, and others in which the entire capillary was prefilled with sample + ampholyte mixture.

Partial Capillary Injection
In the first approach described by Thormann et al. [296], 75 µm i.d. × 90 cm uncoated capillaries were filled with catholyte (20mM NaOH + 0.0–0.06% hydroxypropylmethylcellulose (HPMC)) and a 5-cm segment of sample in 2.5–5% ampholytes was injected at the inlet (anodic end) of the capillary by gravity. After immersion of the capillary inlet in anolyte (10 mM H_3PO_4), high voltage was applied at a field strength of 220 V/cm. Formation of the pH gradient and focusing of proteins into zones occurred as the sample segment was swept toward the detection point at the distal end of the capillary. The addition of HPMC to the catholyte served to dynamically coat the fused silica wall, thereby reducing protein adsorption and EOF. Successful application of this technique depends upon optimization of the HPMC concentration, ampholyte concentration, and sample load to minimize

protein adsorption and to modulate EOF level so that focusing approaches completion before the detection point is reached.

Optimal separation of proteins spanning the whole pH range is difficult with EOF driven CIEF. As for all variations of IEF, salt concentration (higher than 10 mM, in this case) greatly diminishes resolution. Molteni and Thormann [300] studied the factors that influence the performance of CIEF with EOF displacement. The amount of methylcellulose (MC) or hydroxymethylcellulose (HPMC) was found to be optimal in the 0.06–0.1% range. The polymer improved resolution when added to the catholyte and sample solutions, but did not show any benefit if added to the anolyte. Lower polymer concentration produced better resolution of transferrin isoforms.

The concentration of NaOH (catholyte) had a significant effect on migration times of the protein zones, mainly by affecting the rate of electroosmosis. Longer analysis times were observed at higher concentrations of NaOH with an improvement in resolution. On the other hand, higher concentrations of anolyte (phosphoric acid) shortened the analysis time and diminished resolution. Migration was also affected by the concentration of the ampholytes used (1, 2.5, and 5%), with slower mobilization at lower concentrations (1%). Resolution of the transferrin isoforms was decreased when increasing the voltage from ca. 150 to 285 V/cm. The initial length of capillary occupied by the sample was also found to affect migration times and resolution, with longer sample zones providing better resolution at the expense of analysis time.

Since electroosmosis plays a key role in this approach, reproducibility was achieved only through extensive capillary conditioning. A rinse sequence of 0.1 M NaOH, water, and catholyte (5 min each) was recommended. Huang et al. [294] compared the linearity of elution time and pI of model polypeptides in coated and uncoated capillaries. Better linearity was obtained with pressure mobilization in coated capillaries than in uncoated capillaries with EOF driven transport.

Full Capillary Injection

The second method, in which the entire length of the capillary was filled with ampholyte + sample, was described by Mazzeo and Krull

[287,297]. In initial studies using uncoated capillaries, methylcellulose was added to modulate EOF and TEMED was used to block the detector-distal capillary segment. This approach was successful only for neutral and basic proteins due to variations in the rate of EOF during the separation. As the separation progressed, the drop in average pH due to mobilization of the basic segment of the pH gradient into the catholyte resulted in diminished EOF. This, in turn, caused peak broadening and poor resolution for acidic proteins. Improved mobilization of acidic proteins was achieved using commercial C8-coated capillaries (Supelco, Inc., Bellefonte, PA) in which EOF varied less with pH [301]. However, pH-dependent variation of EOF was still significant enough that plots of pI vs. migration time were not linear over broad pH ranges [302]. Use of multiple internal standards was recommended for accurate pI determination with this method.

The main drawback of using uncoated capillaries is that EOF is too high and pH- dependent. Electroosmotic flow normally decreases during the analysis, resulting in nonlinear plots of migration time and pI [302]. The linearity of the plot improves when using coated capillaries. In this case, focusing occurs in the short end of the column, and the majority of the capillary is "blocked" using NaOH or TEMED. The polarity is reversed because EOF occurs from anode to cathode (so the anode is at the capillary end closest to the detector). As can be observed in Figure 12, no focusing peaks are present, which means that only the bands formed at one of the ends of the capillary are being detected. Still, the plot of pI vs. migration time is not linear (Figure 9C). The main advantage of this method is its robustness for the analysis of proteins with moderate pI values.

Single-step CIEF was performed in a capillary first derivatized with octadecyltrichloro silane, and then dynamically coated with various surfactants (Brij 35, PF-108, methylcellulose, polyvinyl alcohol, or polyvinyl pyrrolidone) [303,304]. Copolymers of hydrophilic and hydrophobic monomers interact with the hydrophobic octadecyl layer through the hydrophobic portion of the chain, leaving the hydrophilic segment exposed to the aqueous phase. Several volumes of polymer–surfactant solution are introduced into the column to cover the capillary wall, and then added to the sample +

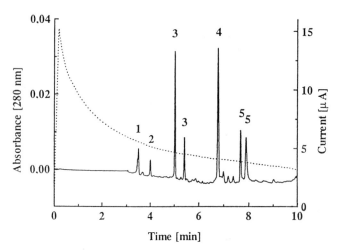

Figure 12 Capillary IEF of standard proteins using residual EOF mobilization. Conditions and peak identities identical to Figure 9C. (Reproduced from Ref. 302 with permission.)

ampholyte solution to maintain a dynamic equilibrium between the wall and the solution. Polymers with higher molecular weights reduce electroosmosis more effectively than surfactants with smaller chains.

Single-Step CIEF in the Presence of Hydraulic Forces

Dynamic CIEF has been reported using siphoning [295]. This force can be applied from the beginning of the analysis (injecting a plug of sample instead of filling up the capillary) to resolve the sample components by dynamic (single-step) CIEF; i.e., the focusing and mobilization step are combined in one. In single-step dynamic CIEF, the sample was injected at 5 psi without filling up the capillary, the inlet reservoir contained 600 μl of H_3PO_4, and the outlet reservoir contained 100 μl of NaOH. Following sample injection, voltage was applied and while the protein zones focused, they were transported to the detection point by the hydraulic flow generated by the liquid height difference of the reservoirs (in this case, approximately 2 cm). In

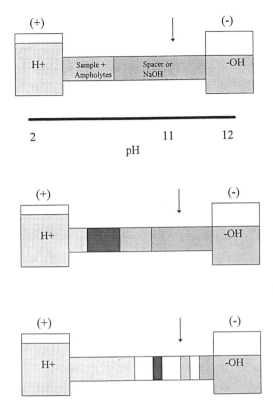

Figure 13 Schematic representation of single-step CIEF in the presence of hydraulic forces. Before focusing starts, the sample occupies a portion of the capillary. For best results the length of the injection zone has to be optimized, while the rest of the column is filled with a spacer or catholyte. A force is applied from the start of the analysis. Proteins focus as they are transported to the detector (arrow).

contrast to pressure or vacuum mobilization, gravity mobilization is the simplest approach, particularly for homemade systems. Single-step CIEF using only hydraulic forces requires the elimination of EOF; otherwise the transport of the zones would also occur because of electroosmosis.

Figure 13 is a schematic representation of single-step CIEF, and Figure 14 represents a typical electropherogram obtained using this

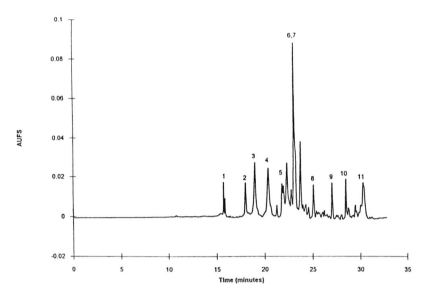

Figure 14 Dynamic CIEF of protein standards. Conditions are identical to those used in Figure 11A, except that the sample occupied only a section of the capillary (sample injected by pressure, 40 psi*s) and the applied force was present from the beginning of the analysis. The detector signal was reduced to half (because the capillary was only partially loaded with sample) and loss of resolution is evident (especially around protein peak number 5). Notice the absence of focusing peaks, and the loss of resolution of the phycocyanin bands. (Reproduced from Ref. 4 with permission.)

method (notice the absence of focusing peaks). The disadvantage of dynamic CIEF is that the capillary is not filled completely and, therefore, the detection signal is reduced due to the smaller amount of sample. The main advantage of this technique is its simplicity.

CAPILLARY SELECTION

To obtain good resolution and reproducibility when performing CIEF with chemical mobilization, it is essential to reduce electroosmotic flow (EOF) to a very low level. In the presence of significant levels of EOF, attainment of stable focused zones is prevented, resulting in band broadening and, in some instances, multiple peaks (Figure 5) caused by incomplete fusion of the nascent zones formed at both capillary ends. Therefore, the use of coated capillaries is necessary for this technique. A viscous polymeric coating is recommended for greatest reduction in EOF, and the use of neutral, hydrophilic coating materials reduces protein-wall interactions. Both adsorbed and covalent coatings have been used for CIEF, but covalent coatings have the advantage of enhanced stability. The most commonly employed coating chemistry has been that described by Hjertén [89]. In this procedure, a bifunctional silane such as γ-methacrylopropyltrimethoxysilane is reacted with silanol groups on the internal surface of the capillary. After covalent attachment of this reagent, the acryl group is reacted with acrylamide in the presence of TEMED and ammonium persulfate without any cross-linking agent to form a monolayer coating of linear polyacrylamide covalently attached to the surface. Capillaries coated with this type of procedure exhibit more than a 40-fold reduction in EOF [305]. It has been reported [97] that the quality of fused silica tubing varies among suppliers and from batch to batch, and that attaining a stable coating requires optimized capillary pretreatment and coating procedures. Capillaries coated with polyacrylamide and other material are commercially available from several sources (see Table 2 in Chapter 4).

The low level of EOF in coated capillaries permits separations to be carried out with very short effective capillary lengths. Earlier work using chemical mobilization was performed using capillaries as short as 11 cm with internal diameters up to 200 μm [293]. More recently, 12–24 cm capillaries with internal diameters of 25 or 50 μm have been used [298]. Theoretically, the resolution should be independent of capillary length in CIEF, because the number of ampholyte species is equal in both cases, only the amount is changed. In practice, however, resolution is diminished with very short capillaries or small sample injections (also when sample and/or ampholytes are very diluted), particularly in single-step CIEF. Since analysis time increases with capillary length, problems associated with protein precipitation are more severe in longer capillaries.

While use of small i.d. capillaries improves resolution through better heat dissipation, the high surface-to-volume ratio increases the potential for protein-wall interactions. These effects are greatly diminished by the use of an inert capillary coating [89]. An advantage of using larger i.d. capillaries is that they are less likely to plug in case of protein precipitation. Although the use of small i.d. capillaries lowers detection sensitivity by reduction in the detector light path, this is generally not a concern since zone concentrations in CIEF are usually extraordinarily high. For example, focusing of a protein into a 0.5–1 mm zone results in a 170- to 340-fold increase in protein concentration relative to the sample. For this reason CIEF can be considered as a potential micropreparative technique. The use of larger i.d. tubes causes the electrical current to increase, but unlike other modes of CE, CIEF can be easily performed in these capillaries due to the current decrease as the run progresses.

The length of capillaries used for single step CIEF is very important, specially when EOF is the driving force. The capillary length has to be optimized according to the size of the injection and the velocity of EOF (or flow due to hydraulic forces), so that the sample mixture will not reach the detection point before it has finished focusing.

CIEF has also been demonstrated in capillaries with the wall chemically modified by the attachment of a hydrophobic agent. These capillaries present a high level of EOF when the pH is above 2–3 but, surprisingly, they exhibit reduced EOF when performing CIEF, perhaps due to ampholyte-wall interactions. For an unknown reason, not all capillaries with hydrophobic coating are suitable for CIEF, and mixed results can be experienced with multiple capillaries having the same chemical modification.

The inner surface of capillaries can be chemically modified to eliminate electroosmosis and to reduce protein adsorption to the capillary wall. The lifetime of the capillary coating is dictated by silica chemistry, polymer type, and polymer stability. Silica has been used as a chromatographic support for several decades, and it is well known that the stability of bonded phases decreases with increasing pH, but stability is still quite satisfactory at pH 8.0–8.5. The most typical CIEF set up incorporates the use of a "spacer" added to the ampholytes, which usually span a pH range from 3 to 10. Under these conditions, approximately 30–40% of a 17–24 cm capillary is filled with a solution of a pH higher than 8.0. In a survey of the isoelectric points of approx. 2380 polypeptides [306], the pI distribution is as follows: 89.66% with a pI below 8.0; 4.12% with a pI between 8.0 and 8.5; and 6.22% with pI above 8.5. This suggests that a CIEF system that extends from a pH of 3–8.5 will be useful to the analysis of approximately 94% of all proteins. This is of significance because at pH 8.5, a high number of runs can be achieved in a single capillary. Figure 15A shows two superimposed CIEF analyses of transferrin in an ampholyte pH range spanning from 3.0 to 8.5. A spacer with a pI of 8.7 was used to block the blind side of the capillary. Figure 15B shows run number 500 in the same capillary under identical conditions [307].

DETECTION

UV/Vis Absorption

Most applications published to date employ on-line detection of mobilized proteins by absorption in the ultraviolet or visible spectrum at a fixed point along the capillary. The strong absorbance of the ampholytes at wavelengths below 240 nm makes detection of proteins in the low UV region impractical. Therefore, 280 nm is generally used for absorbance detection in CIEF. This results in a loss in detector signal of as much as 50-fold relative to detection at 200 nm (see Table 1 in chapter 3), but the high protein concentrations in focused peaks more than compensate for the loss of sensitivity imposed by 280 nm detection. Since ampholytes may still be detected at 280 nm, care should be exercised when analyzing diluted or low absorbing proteins [308]. Since ampholyte species vary in abundance and UV extinction, the lowUV profile of ampholytes detected during CIEF can provide some information about the ampholyte distribution in the capillary. When a multiwavelength or scanning UV detector is used, correlation of the migration position of the ampholyte peaks with protein standards could allow the ampholyte profile to be used for internal standardization. In some instances, proteins possess chromophores that can be detected in the visible range of the spectrum, e.g., hemoglobin and cytochromes, or detection by laser induce fluorescence of tagged antibodies directed against the protein of interest can be used [39].

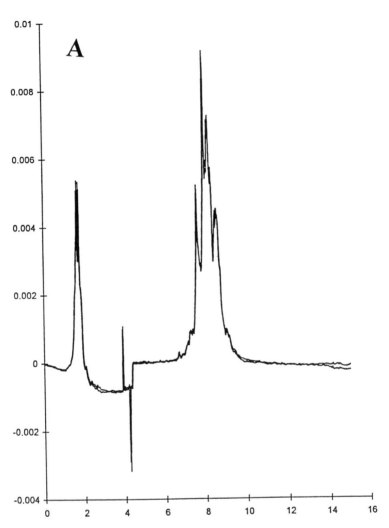

Figure 15 Capillary IEF analysis of transferrin in a 3.0 to 8.5 ampholyte gradient. A spacer with a pI of 8.7 was added to the ampholytes to block the blind segment of the capillary. Analysis conditions: 24 cm × 25 μm capillary coated with linear polyacrylamide; focusing at 15 kV for 4 min; ion-addition mobilization at 15 kV; on-line detection at 280 nm. All solutions and the capillary were thermostated at 20°C. (A) overlay of runs 16 + 17; (B) run 500.

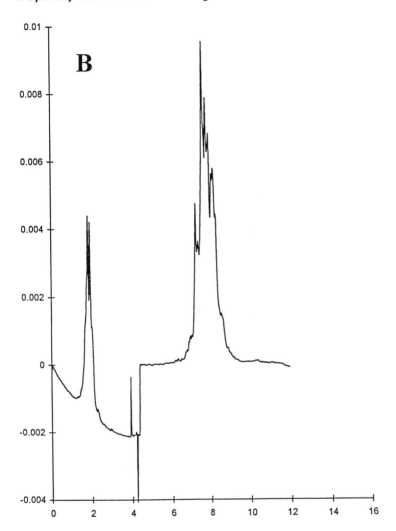

Focused zones can also be detected by sliding the capillary in front of a modified detector [310]. In effect, the entire capillary is mobilized instead of its contents. Since most capillaries used in CE are externally coated, a UV-transparent externally coated capillary or a capillary with the external cladding removed with hot sulfuric acid is used. To slide the capillary, a string is attached to an electrical motor and to the column. Two related problems may interfere with detection: noise due to friction, and noise due to vibration. The main advantages of this technique are a uniform rate of mobilization, the capability of performing signal averaging, and shorter analysis time. Although the main purpose of this approach is not to disrupt the equilibrium attained by the system at the completion of focusing, this advantage is compromised if mobilization is carried out with the voltage turned off, thus allowing the protein zones to diffuse.

Concentration Gradient Detection

The use of concentration gradient detectors with various formats have been reported extensively [310–319]. The system incorporates a capillary mounted in a holder that aligns the column to a HeNe laser beam. A positioning sensor is located at the exit side of the laser beam, and it detects deflections generated by the passage of substances with a refractive index different than that of the background buffer. The main advantage of this detector is its universality. Because it is a universal detector, ampholytes may produce signals during the mobilization step. The derivative nature of the detector enables recognition of the sharp bands generated by the protein zones against the background of the broader zones produced by the ampholytes. The detection system can also be built to scan along the capillary, performing detection of focused protein zones without mobilization. Optimization produces fast analysis times (2 min) and detection limits in the 1–5 mg/ml range. Use of capillary arrays greatly improved throughput. Using this system peptides produced by the tryptic digestion of bovine and chicken cytochrome C were analyzed. The main significance of this application is that not all peptides contained aromatic amino acids, which are required for UV absorption detection using 280 nm.

Detection using a diode array detector resulted in noisy baseline, and decreased resolution. This phenomena was consistent with all three types of mobilization used, and with different ampholytes [302].

Using modes of detection that require chemical modification (derivatization) of the sample molecules can change the pI, often produce multiple peaks, and therefore, are not widely used.

Laser-Induced Florescence

Laser-induced fluorescence is a highly sensitive mode of detection, but lasers that emit at a visible wavelength often require derivatization of the sample prior to analysis. This drawback is eliminated by using a laser that emits in the UV range. Laser-induced native florescence at 275.4 nm from an argon–ion laser is so highly sensitive that it allows the detection of hemoglobins from a single cell [320]. The injection volume from a single red cell was estimated to be less than 90 fL. Hemoglobin analysis from single red cells was used to determine variants in normal, diabetic, fetal, and sickle adult erythrocytes. Single-step CIEF was performed in a 40 cm total length (30 cm effective length) uncoated capillary with an i.d. of 21 μm in combination with 0.1% methylcellulose (25 cp) to reduce EOF. The ampholyte solution consisted of 0.5% Ampholine (pH range 5–8) containing 0.1% MC. The concentrations of the ampholytes, anolyte, and catholyte were selected in order to reduce interference, and thus, to improve detection. Hemoglobin A_0 standard was injected by gravity (11 cm height differential for 10 s) and it was estimated to introduce a sample volume of 0.14 nl. Individual cells were injected by using a syringe: the outlet reservoir was sealed with a septum, then the syringe needle was introduced through the septum. The inlet end of the capillary was immersed in 200 μl of 8% sucrose to which 10 μl of cell suspension were then added (the sucrose solution was used to desalt the cell suspension). By gently drawing the syringe plunger, a vacuum was created and a cell introduced into the column. The presence of a single cell was confirmed by visually inspecting the capillary end under a microscope. The focusing/mobilization step was driven by a constant voltage at 24 kV, using 1 mM phosphoric acid as anolyte, and 2 mM NaOH as catholyte. The 275.4 nm line of an argon–ion laser was isolated with a prism and focused onto the capillary. This wavelength was used as the excitation source. Fluorescence was collected with a microscope objective and passed through two color filters before reaching the photomultiplier. Peak identity was based on migration times and the known composition of the samples. Linear regression ($r^2 = 0.9984$) analysis generated pI values for Hbs A_0, A_{1c}, S, and F identical to those found in the literature. Figures 16A and 16B show the analysis of normal and heterozygous sickle cell single red cells.

Capillary IEF and Mass Spectrometry

The separation power of CIEF often generates a high number of peaks even when relatively pure samples are analyzed. For most applications, the next stage of the process is to characterize and identify the sample components observed. As already discussed, one of the advantages of CIEF is its potential micropreparative capabilities. Capillary IEF allows the collection of fractions that can be further analyzed by other methods. Some of the most widely used characterization tools include mass spectrometry, peptide mapping, and amino acid analysis.

Foret et al. [321] collected fractions of model proteins and variants of human hemoglobins after fractionation by CIEF, and then analyzed them by matrix-assisted laser desorption time-of-flight mass spectrometry (MALDI-TOF-MS). As the authors point out, MS is an orthogonal method to CIEF since it separates according to molecular mass. The anolyte consisted of acetic acid (pH 2.9) and the catholyte was ammonium hydroxide (pH 10.9). After focusing (2 min at 10kV) in a coated capillary (75 μm × 40 cm), fractions were collected during mobilization, which was accomplished by raising the injection end of the capillary 5 cm (8 cm for hemoglobins). Under these conditions, the hydrodynamic flow obtained was 22 nl per minute (5 cm height differential), or 35 nl per minute (8 cm height differential). The flow was calculated using fluorescein. The homemade instrument was equipped with a sheath flow interface (equipped with a syringe pump that supplied ammonium hydroxide solution) coupled to 60 collection capillaries (with a volume of 20 μl each), which were positioned at the outlet of the separation capillary by a computer controlled stepper motor. An optical fiber was positioned 1 cm from the capillary's outlet end, which served to detect the migrating zones and allowed precise calculation of the times to change the

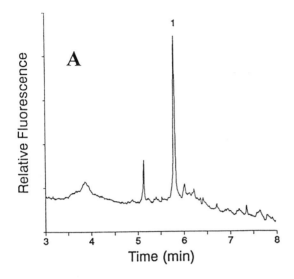

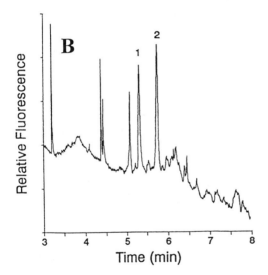

Figure 16 Hemoglobin analysis from a single cell using laser-induced native fluorescence detection. (A) Analysis of a normal adult erythrocyte homozygous for HbA. (B) Analysis of a heterozygous sickle red cell. Notice that the concentrations of Hb S (peak 1) and Hb A (peak 2) in the sickle cell are fairly similar (44.7 and 55.3%, respectively). Peak identification was based on migration times (Reproduced from Ref. 320 with permission.)

collection capillary. This instrumental arrangement allowed main-
tainance of the electric field during the entire analysis, and because
the detection point was very close to the capillary's end, the use of a
blocking spacer was eliminated. After collection (approx. 13 μl per
fraction), the standard proteins were directly deposited on the probe
and allowed to dry, then sinapinic acid was added to the sample probe.
Hemoglobins were dried and frozen (−20°C) before being redissolved
in the MS sample matrix. The mass spectra results for the model
proteins (myoglobin, β-lactoglobulin A, and carbonic anhydrase I
and II; these proteins differ in both molecular weight and isoelectric
point) correlated well with previously reported molecular weights,
with an estimated mass error of less than 1%. The pIs of the same
sample were estimated to within 0.1 pH unit. The hemoglobin sample
(mixture of hemoglobins A, F, S, and C) was easily resolved by CIEF
(these proteins differ in pIs, but the mass difference is very small),
but produced two bands when analyzed by MS without performing
CIEF. As expected, MS of the collected fractions produced nearly
identical traces for every hemoglobin variant.

OPTIMIZING CIEF SEPARATIONS

As stated in the introduction of this chapter, CIEF is a member of a
group of techniques characterized by the attainment of equilibrium
as the end point of the analysis. Resolution in these methods is
achieved by the creation of a gradient, where sample components are
transported by a force (electrical, gravity, etc.) until each of the
analytes reach a point of equilibrium, where the force of transport
equals a force of resistance to motion (e.g., isoelectric point, density).

Resolution in IEF

According to Righetti [282], resolution in IEF can be calculated using the following equation:

$$\Delta(\text{pI}) = 3 \left[\frac{D(d\text{pH}/dx)}{E(-d\mu/d\text{pH})} \right]^{1/2} \tag{6-4}$$

Where D is the diffusion coefficient of the sample, $d\text{pH}/dx$ is the rate
of change of pH with distance x (i.e., the slope of the pH gradient), E

is the applied electric field, and $-d\mu/d$pH is the change in mobility of the sample with pH. Experimentally, the variables that can easily be manipulated are the electric field (higher voltages leading to increased resolution), and dpH/dx with narrower pH gradients leading to higher resolution. The estimated maximum resolving power of IEF is 0.02 pH unit when carrier ampholytes are used to create the pH gradient [282]. The two variables intrinsic to the sample molecules are the diffusion coefficient and the rate of change of mobility with pH near the isoelectric point. It is desirable that the analyte has a low diffusion coefficient and a large change of mobility with pH. In general, proteins satisfy this requirements fairly well, but smaller peptides do not, since they have larger diffusion coefficients and smaller changes in mobility with pH. To obtain narrow protein zones, the rate of refocusing must be greater than the rate of diffusion. This can be accomplished by performing the separations in high viscosity media to reduce diffusion or by using high electric fields.

As previously mentioned, resolution in IEF strongly depends on the ampholyte composition. The Law of Monotony [282] formulated by Svensson in 1967 states that a natural pH gradient increases continually and monotonically from the anode to the cathode, that the steady state does not allow for reversal of pH at any position along the gradient, and that two ampholytes (in stationary electrolysis) cannot be completely separated from one another unless the system contains a third ampholyte of intermediate pH (or pI). The latter explains why better resolution is obtained when mixing ampholytes from different vendors and production batches: as the number of ampholytes species increases, the chance that one or more ampholytes have intermediate pI than those of the sample components also increases.

Parameters That Affect the Performance of CIEF

The resolving power of equilibrium methods is very high, and usually there are only a few parameters that can be modified to increase resolution. The most important of these is the manipulation of the slope of the gradient. Two components will be at a greater distance one from the other in direct relation to the shallowness of the gradient.

Since the pH gradient in CIEF depends in the quality of the ampholytes, greater separation power can be achieved by increasing the number of species present in a particular range. One of the simplest strategies to achieve this is to mix ampholytes, specially from multiple vendors. Manipulation of the pH slope created by the ampholytes is usually the only parameter that allows more resolution when CIEF is totally optimized, but there are other factors that affect resolution and minimizing their effect can greatly improve results.

Effect of Salt

The single most important parameter to control in the sample composition is the concentration of salt (including simple salts and buffer components). The negative effect of salt is compression of the pH gradient, which increases the gradient slope, reduces resolution, increases the risk of precipitation, and generally slows the focusing and mobilization steps.

Gradient compression can be explained by the following mechanism: Whenever an ion migrates out of the capillary because of its electrophoretic mobility, it must be replaced by an ion(s) of equal charge from the anolyte or catholyte reservoir. Thus positively charged ions are replaced by protons from the phosphoric acid reservoir, and negatively charged ions are replaced by hydroxyls from the sodium hydroxide. Ampholytes and proteins do not migrate out of the capillary because at some point inside the capillary they reach their isoelectric point and stop, and they do not have to be replaced by outside ions. But salts do not have an isoelectric point, and when they exit they cause a "rush" of protons and hydroxyls that effectively reduce the volume of the capillary where the ampholytes can create a pH gradient (obviously no ampholyte can focus in the area of extreme pH). This is illustrated in Figure 17. This phenomena also has a deleterious effect on the lifetime of the coated capillaries, since the coating is exposed to extreme pH. One possible way to reduce the effect of salts, is to increase the capillary length, e.g., if a gradient is compressed to half the capillary length, doubling the capillary length should "stretch" it back to its original size. To keep the amount of sample constant, the sample could be diluted to half, effectively reducing the salt concentration. The drawbacks include a shortened

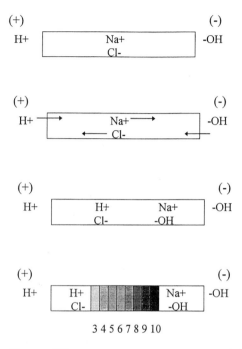

Figure 17   Schematic representation of the effect of focusing in the presence of salt. Salt ions (Na^+, Cl^-) exiting the capillary under the influence of an electric field are replaced by ions of the same charge from the reservoirs (H^+ and OH^-). The presence of salt inside the capillary and the application of an electric field results in the formation of NaOH and HCl at the capillary ends (top three panels, depicted in the absence of ampholytes). This rapid influx of protons and hydroxyls compresses the pH gradient (bottom), reducing the effective length of capillary, diminishing resolution, and increasing the risk of precipitation.

capillary lifetime (exposure of capillary ends to extreme pH) and sample-to-sample variations (since gradient compression is proportional to the amount of salt present in each sample).

Even small variations in the salt content of the sample can introduce significant error in the migration time of the protein zones [307]. Figure 18 displays four different analyses of human hemoglobins A, F, S, and C that differed only in the salt added to the sample mixture

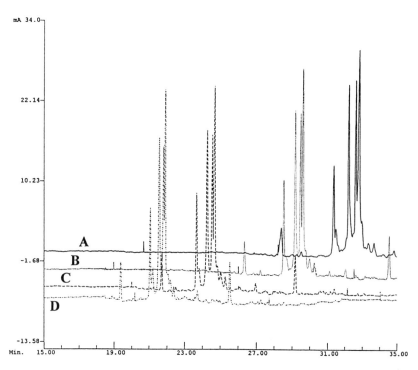

Figure 18 Migration time shift due to the presence of various amounts of salt present in the sample matrix. A standard mixture of human hemoglobins A, F, S, and C were spiked with (A) 40, (B) 30, (C) 20, and (D) 10 mM Tris-Phosphate pH 8.0. Proteins focused in the presence of high salt eluted earlier (with diminished resolution) than samples with low salt concentration. Analyses were carried out in 24cm × 25 μm capillaries coated with linear polyacrylamide. Focusing at 15 kv for 240 s, ion-addition mobilization at 15 kV. Capillary and samples were thermostated at 20°C. (Reprinted from Ref. 4 with permission.)

(10, 20, 30, and 40 mM Tris phosphate, pH 8.0). In this case, the buffer used to increase the salt content of the sample accelerates the elution of the protein zones, but also has adverse effects on resolution. If migration time from this data is used for peak identity the error introduced by varying amounts of salt is significant, whereas the identification of protein zones by their pI (obtained by "bracketing" the sample with pI markers) is much more accurate (Table 1) [307,322].

Table 1A Migration Time Precision with Increasing Sample Salt Concentration

Salt Concentration	Hb C	Hb S	Hb F	Hb A
0 mM	15.48	16.82	17.41	17.57
20 mM	15.34	15.77	15.98	16.04
30 mM	14.86	15.2	15.37	15.4
40 mM	14.51	14.79	14.95	14.97
50 mM	14.15	14.7	14.83	14.85
Average	14.93	15.46	15.71	15.77
Std. Dev.	0.47	0.87	1.05	1.11
%RSD	3.15	5.64	6.7	7.05

Table 1B Isoelectric Point Precision with Increasing Sample Salt Concentration

Salt Concentration	Hb C	Hb S	Hb F	Hb A
0 mM	7.34	7.01	6.88	6.88
20 mM	7.38	7.10	6.97	6.93
30 mM	7.39	7.12	6.97	6.94
40 mM	7.40	7.11	6.96	6.94
50 mM	7.39	7.11	6.97	6.95
Average	7.38	7.09	9.95	6.92
Std. Dev.	0.02	0.05	0.04	0.05
%RSD	0.32	0.64	0.57	0.65

Strategies to Decrease Salt Concentration

1. Dilution. Unlike other forms of capillary electrophoresis, the effect of salt concentration in CIEF can be minimized by simply diluting the sample with the ampholyte solution, up to the point where detection becomes a limiting factor. The ampholyte solution should be maintained constant by using 50% v/v of a prepared 2X solution. Samples should then be diluted with the ampholyte solution (and deionized water if necessary) to avoid conductivity problems.

2. Gel Filtration. Gel filtration can be employed for proteins and large peptides, usually with a gel of very small pores, so

that sample components are not retained. It is advisable to desalt the sample prior to the addition of the ampholytes.

3. Ion-Ampholyte Replacement and Electric Field Ramping. Another option is to add an on-line desalting step to the CIEF process [323] prior to sample focusing. In this approach, the sample and ampholytes are injected into the capillary. A solution of ampholytes is titrated to pH 4.0 with HCl, and it is used as anolyte. Another solution is titrated with NaOH to pH 11.0, and is used as catholyte. An electric field employing constant current (e.g., 10–40 μA) to minimize heating is applied. Under this condition, the ampholytes placed in the reservoirs possess a net charge, and salt ions exiting the capillary can be replaced by a combination of ampholytes and protons (anode) or ampholytes and hydroxyls (cathode), resulting in a reduction of pH gradient compression (because the influx of ions into the capillary is a combination of protons + ampholytes, and hydroxyls + ampholytes, gradient compression is probably not completely eliminated). The electric field is maintained until the voltage reaches a determined value, e.g., 3–10 kV, then turned off. The CIEF process then continues as a regular CIEF analysis using conventional anolyte, catholyte, and mobilization reagents. This routine has been used to desalt samples containing up to 0.5 M NaCl (Figure 19). This method is advantageous only when a small quantity of sample is available. The main disadvantages are the increase in analysis time, and variations introduced by the continuous addition of the ampholytes (samples containing different amounts of salt will be focused with correspondingly different final concentration of ampholytes). Another point of concern is the integrity of the pH gradient linearity; since ampholytes migrate at different speeds, those with higher mobility will enter the capillary in correspondingly higher concentrations, thus changing the distribution of their concentration along the capillary. This procedure requires the suppression of electroosmosis.

As already stated, samples containing increased amounts of salt generate a proportionally higher electric current (and therefore Joule heat). Coupled with an uneven distribution of the electric field, proteins could denature (often leading to precipitation) and the risk of bubble formation is increased, resulting in breakage of the electric current. This problem can be minimized by slowly increasing the strength of the electric field, either by partially performing the analysis at constant current [323], or by voltage ramping [324]. Voltage ramping as described by Clarke et al. [324] employed a 37 cm × 50 μm PVA-coated capillary filled with a mixture of sample and ampholytes. Prefocusing ("desalting") was performed using a voltage ramp from 0 to 10 kV over 6 min, then the voltage was maintained at 10 kV for 2 min before a rapid focusing ramp to 20 kV over 30 s, and maintained over 5 min. The anolyte (50:49:1 v/v/v methanol/water/acetic acid) and catholyte (50:49:1 v/v/v methanol/water/ammonium hydroxide) were prepared to allow future MS analysis. As previously described, when focusing of high salt samples occurs in the presence of anolyte (acid) and catholyte (base), gradient compression results in loss of resolution, and decreased capillary lifetime. In this protocol, perhaps the length of the capillary compensates for loss of resolution.

4. Dialysis Fibers. To remove salts from the sample before the performance of CIEF [325], a cartridge with an attached dialysis fiber has been used. The dialysis fiber was immersed in a beaker containing a 4% ampholyte solution. One end of the fiber was attached to the electrophoresis column, whereas the other was used to introduce the sample. Since the molecular-weight cutoff of the fiber could be manipulated (e.g., 9 kD), it served a dual purpose: while salt ions diffused out, ampholytes diffused in. Due to the narrow diameter of the fiber (75 μm) desalting was accomplished in about 1 min. A special device was necessary to interface the desalting system to the capillary.

5. Dialysis. Dialyzing against the ampholyte solution is an effective way to eliminate undesirable salts. The main drawbacks of this approach are the time required for the salts to diffuse out of the sample compartment and into the bulk of the dialysis solution, and the cost of large volumes of

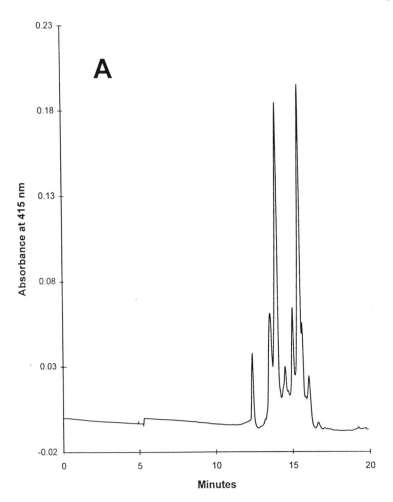

Figure 19 Desalting using ion–ampholyte replacement. A sample of patient blood containing hemoglobins A and C was diluted 1:200 in 2% Bio-Lytes 3–10. Separations were performed in a 24 cm × 50 μm polyAAEE-coated capillary thermostatted at 20°C; detection was at 415 nm. The capillary was filled with sample by pressure injection for 40 s at 100 psi. Panel A: Control separation of diluted blood sample. Focusing was performed at 15 kV for 300 s with 20 mM phosphoric acid as anolyte and 40 mM NaOH as catholyte. Ion-addition mobilization was performed at 15 kV using 20 mM phosphoric acid as anolyte and a zwitterion solution as catholyte. Panel B: Sample was spiked with NaCl at a final concentration of 250 mM and

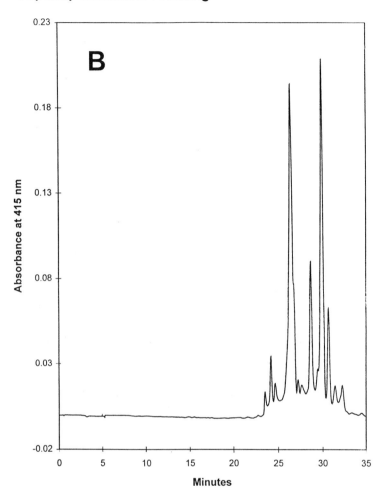

desalted on-line by ion–ampholyte replacement. After injection, desalting was performed using 2% BioLytes 3–10 (adjusted to pH 4.0 with HCl) as anolyte and 2% BioLytes 3–10 (adjusted to pH 11.0 with NaOH) as catholyte. Desalting was performed at 24 μA constant current; when the voltage reached 15 kV, the desalting step was terminated and isoelectric focusing was performed as described for panel B. Note that the resolution of the desalted sample is better than that of the unspiked sample; this is probably due to gradient compression by naturally occuring ions in the use of unspiked blood sample. Analysis of the spiked sample without desalting was impossible due to extremely high current (>300 μA) at the onset of focusing.

ampholyte solution required. In our experience, this process takes a minimum of 2 h, but sometimes satisfactory results are reached only after dialyzing overnight (if possible in a cold room). To determine the optimal time of dialysis, the amount of sample and the membrane surface area must be taken into account. Membranes with a molecular-weight cutoff of 3000 Da are readily available from several sources. The volume of the solution used should be at least 50 times greater than that of the sample. Since CE requires small amounts of sample, a suggested dialysis method employs a floating dialysis membrane carefully placed on top of the bulk solution. Due to surface tension, the membrane floats on the liquid surface. Small quantities of sample (usually 50 μl or less) are placed on the membrane and allowed to dialyze. Using this approach, samples are dialyzed within an hour (due to the short diffusion path). The main disadvantage of this method is the occasional sinking of the membrane, which needless to say, results in total sample loss. We prefer the use of small dialyzing devices, which are available commercially from several sources (e.g., Pierce).

6. Ultrafiltration. A more rapid approach for desalting is ultrafiltration. In this technique, the sample solution is placed in the upper compartment of a microcentrifuge vial whose bottom is a membrane with a specific molecular weight cutoff (e.g., 3000 Da). The lower compartment of the vial is empty, and serves as a reservoir for the solvent that crosses the membrane. Due to membrane pore size variations, most vendors recommend the use of a molecular-weight cutoff at least 3 times smaller than the size of the sample components of interest. The vial containing the sample is placed in a microcentrifuge, and centrifuged for several minutes (20–45 min, if possible in a cold room or under refrigeration) at full speed. The centrifugal force drives molecules toward the bottom of the vial, but species of molecular weight higher than the size of the membrane pores are retained. This process not only eliminates the amount of salt present, but can also concentrate the protein by recovering the sample with a smaller vol-

ume of solvent. For quantitative analyses, dilution factors and recovery from the membrane have to be estimated. The main drawback of ultrafiltration is the cost of the vials (which may significantly increase the cost per analysis) and the time required to perform the process.

Additives and Protein Precipitation

Since a major problem in IEF is the precipitation of sample components the most common type of additives used in CIEF are solubilizing agents. Protein precipitation causes irreproducibility in migration times, peak area (thus affecting quantitation), poor pattern reproducibility, capillary clogging, unstable current, slow mobilization, and other undesirable effects. Protein precipitation, and additives used to prevent it, are treated in more detail in the section dealing with protein precipitation below. Some common additives in CIEF include:

Hydrophilic Polymers

Hydrophilic polymers are used frequently in CIEF to reduce EOF, either alone or in combination with capillary coatings. Various amounts of hydrophilic polymers are added to sample and/or electrolyte solutions during hydraulic mobilization. Other uses of polymeric additives include:

1. Fluid Stabilization. High-concentration focused protein zones may have a higher viscosity than the surrounding ampholyte solution, thus giving rise to convective currents at the interface of the two solutions. These convection currents may result in decreased resolution, and their effect can be minimized by increasing the viscosity of the solution through the use of hydrophilic polymers [326].
2. Increased Viscosity During Hydraulic Mobilization. Polymer additives are used in almost all modes of hydraulic mobilization. The main function of these polymers is to increase the viscosity of the sample matrix, the anolyte and catholyte solutions. The force applied during hydraulic mobilization must be optimized usually to a very low value, or the viscosity of

the solutions must be increased to avoid zone distortion due to the laminar nature of hydraulic flow. Protein zones focused in low-viscosity media are pushed out of the capillary very fast (resulting in band distortion to the point that proteins are not detected) even if only about 1 psi of pressure is applied to a 24 cm × 50 μm capillary. Methylcellulose has been used to increase the viscosity of the sample + ampholyte solution to slow down the flow velocity during pressure driven mobilization in a neutral, hydrophilic, coated capillary [294]. The concentration of methylcellulose must be optimized, and better results may be obtained with higher concentrations of polymer. Polymer-containing solutions tend to trap air bubbles, which are detrimental to the electrophoretic process (because of their high electrical resistance), and thus these solutions should be degassed (e.g. by centrifugation). High resolution can be obtained using electric fields of about 500 to 900 V/cm. When the applied force is small enough (e.g., during gravity mobilization), the use of polymers is not always necessary [295].

Protein Precipitation

Protein precipitation is a common problem associated with IEF. In fact, isoelectric precipitation is used as a coarse means of protein purification. Just as in slab gel IEF, protein precipitation is a major source of difficulty in capillary isoelectric focusing. Precipitation in CIEF is manifested by current fluctuation and/or loss, by variations in peak heights or migration patterns, and by spikes in the electropherogram generated as particulates (protein aggregates) transit the detection point (see Figure 20).

Possible causes of protein precipitation include:

1. Precipitation Due to Salt Removal. During the focusing process, salt ions (which have no isoelectric point) exit the capillary, leaving the proteins behind. Salts enhance protein solubility by reducing ionic interaction of different polypeptide chains, and thus when the salts are lost, protein interaction increases.

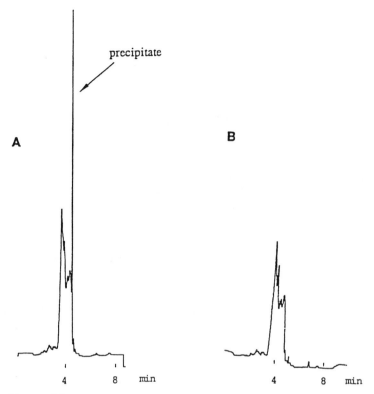

Figure 20 Protein precipitation during CIEF. This analysis of monoclonal antibodies produced extremely narrow peaks (usually irreproducible) due to protein precipitation (A). The addition of detergents such as Triton X-100 prevents the aggregation of proteins (B). Capillary IEF performed in 14 cm × 25 μm capillaries coated with linear polyacrylamide. Sample dissolved in 2% Bio-Lytes 3–10. Focusing at 6kV. Ion-addition mobilization at 8 kV. (Reproduced from Ref. 298 with permission.)

    2.   Precipitation Due to Denaturation and Electrostatic Interactions. The proteins eventually reach their isoelectric point, concentrating in a small area of the column, where, in the absence of salts, they are free to interact with each other (because the proteins are isoelectric electrostatic repulsion is eliminated) forming agglomerates that may eventually precipitate. Some proteins are inherently unstable when they are

isoelectric, they denature during focusing, and the denaturation process may render them insoluble. Proteins with these characteristics are difficult to analyze by IEF.

Strategies to avoid both of the above problems include the reduction of sample concentration, short focusing times, use of chaotropic agents (e.g., urea), and nonionic detergents (Brij 35, Triton X-100, etc.) [298]. The concentration of the ampholytes can be increased to reduce electrostatic interactions. In the presence of a high concentration of ampholytes, it is more likely that the charges of a polypeptide will interact with an ampholyte rather than with another polypeptide, thus reducing the risk of protein aggregate formation. Higher concentrations of ampholytes tend to increase focusing time in general, and also mobilization time when using chemical mobilization. Increased ampholyte concentration (3–6% w/v) works better when combined with decreased sample concentration.

The main principle applied in all of these strategies is to make it difficult for the proteins to find each other, and to substitute protein–protein interactions for protein–additive interactions.

3. Precipitation Due to Hydrophobic Interaction. Protein extracted from hydrophobic environments (e.g., membrane proteins), or proteins with hydrophobic patches exposed on the surface of the molecule (e.g., after denaturation) naturally tend to aggregate. These proteins are often extracted using series of detergents, and if those detergents are nonionic they should be added to the ampholytes + sample solution. Small amounts of organic solvents are used for mild cases of aggregation due to hydrophobic interaction. Ionic detergents will either affect the isoelectric point, make the focusing process impossible, or if unbound to the protein, exit the column upon application of the electric field; they should not be used as solubilizing agents in IEF.

Protein precipitation can be minimized by reducing the focusing time (which can reduce resolution and produce "double peaks"; Figure 5) or by reducing protein concentration (which reduces sensitivity). The most effective means of reducing protein precipitation is addition of protein solubilizing agents such as surfactants or small quantities of organic solvents to the sample + ampholyte mixture. Modifiers such as ethylene glycol (about 5%) serve to reduce hydrophobic interactions which promote aggregation [292]. We have found nonionic detergents such as Triton X-100 to be quite effective in reducing precipitation; however, the reduced form (available from Sigma Chemical Co., St. Louis, MO) should be used whenever possible. Typical concentrations of Triton X-100 used are 0.1–1%. Brij-35 (Fluka Biochemika, Switzerland) is a UV-transparent nonionic detergent which has also proved useful. We have, on occasion, used chaotropic agents such as 4–8 M urea to help solubilize proteins. However, chaotropes and aggressive surfactants can change the migration behavior of proteins. Selection of proper additives must not only reduce the precipitation problem, but be compatible with the aim of the analysis (e.g., chaotropes tend to disrupt tertiary and quaternary structures of polypeptides; see "Denaturing IEF" below). Since precipitation is due to gravitational sedimentation of aggregates, hydrophilic polymers can be used to increase the viscosity of the solution (reducing the speed of sedimentation). Care should be taken in selecting the polymer type and concentration to avoid excessive delays in protein migration.

Denaturing IEF

Poor protein solubility under IEF conditions has limited the number of applications developed for this technique. Solubilization in some cases may be a problem even before the IEF analysis is initiated [327]. Some proteins, e.g., membrane proteins, are difficult to analyze in typical electrophoresis buffers. These polypeptides, however, can be rendered soluble in the presence of additives such as SDS. Unfortunately, SDS or similar ionic detergents cannot be used in IEF, since they eliminate the amphoteric properties of proteins. Protein aggregates form mainly due to hydrophobic interactions and hydrogen bonding. An additive that can be used to increase protein solubility is urea. Unfortunately, hydrophobic interactions and hydrogen bonding play a major role in protein tertiary structure, and because urea disrupts these forces, protein denaturation occurs. When the protein is denatured, all hydrophobic residues are exposed, increasing the possibility of hydrophobic interactions. For this reason, detergents are used in combination with urea. Not all detergents are suitable for denaturing IEF. Besides the need for the detergent to be non-ionic or zwitterionic, when using urea it is preferable to add detergents with branched, cyclic, or short aliphatic chains. This requirement is due to the formation of "channels" by the urea at high concentration. These channels tend to trap linear alkyl chains. Some detergents used with success include Tritons (reduced form), Nonidet P-40, CHAPS, octyl glucoside, and lauryl maltoside. Typical concentrations used range from 0.1 to 5%.

The use of urea must be approached with caution, since urea solutions often contain ammonium cyanate (the concentration of this compound increases with temperature and pH) which can react with the amino group of lysines and the amino terminus leading to artifact peaks. This effect is minimized by the presence of ampholytes, whose primary amines are cyanate scavengers, and by deionizing the urea solution with a mixed-bed resin prior to adding the ampholytes and the detergent.

Figure 21 shows a CIEF analysis of α-casein under denaturing conditions in the presence of 8 M urea and 1% reduced triton. Caseins are notorious for low solubility at their pI, and in fact, isoelectric point precipitation is used to fractionate caseins from milk. In the absence of urea and Triton, a variable number of peaks (or no peaks) was obtained. Denaturing and reducing CIEF in the presence of 8 M urea and 5% β-mercaptoethanol has also been reported [302]. Reduction of protein samples prior to IEF analysis should be approached with caution, because the reduction product often depends on the conditions used. In most cases, heat is used to completely reduce the protein chains, but the high temperatures used (usually a boiling water bath) will denature the polypeptide. Protein denaturation may substantially shift the pI of the molecule and can lead to decreased solubility, whereas incomplete denaturation can result in multiple peaks.

Hydraulic vs Chemical Mobilization

Perhaps because mobilization is viewed as the simple transport of focused zones toward the detector, this stage of CIEF is often over-

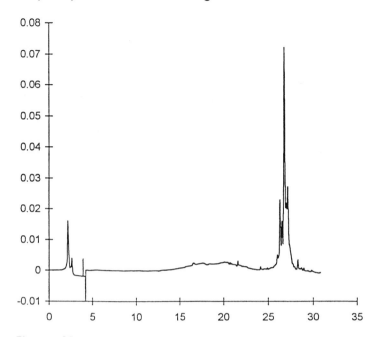

Figure 21 Capillary IEF analysis of α-casein under denaturing conditions. Caseins are notorious for their tendency to precipitate at or near their isoelectric points. The α-casein was dissolved in Pharmalyte 3–10 (5% v/v) + 8 *M* urea + 1% reduced Triton X100. Analysis conditions: 22 cm × 50 μm coated capillary, thermostated at 25°C; Focusing 4 min at 15 kV; ion addition mobilization (zwitterionic) at 15 kV. Detection at 280 nm.

looked during method optimization. Here, we review how hydraulic mobilization can be finely tuned by manipulating the pressure applied, and ion addition mobilization can be greatly improved by properly selecting the type of ion used and its concentration.

Since CIEF is an equilibrium technique, protein focusing is dictated by the analysis conditions used. But in reality resolution can be lost during the mobilization step if the type of mobilization is not carefully chosen, and also if mobilization parameters are not well optimized [307]. Chemical and hydraulic mobilization methods represent alternatives that complement each other. Generally, chemical mobilization generates sharper peaks in the neutral to basic end of the gradient (protein zones focused closer to the detector), whereas hydraulic mobilization increases mobilization efficiency at the far end of the capillary (acidic proteins).

Chemical mobilization offers good linearity of migration time vs. pI plots (linear correlation coefficient of 0.991), and reproducibility comparable to hydraulic mobilization [302]. One advantage of chemical mobilization is that it preserves resolution better than hydraulic mobilization, as demonstrated for the analysis of antibodies.

Besides affecting resolution, the type of mobilization used determines the linearity of pI vs. migration time [302]. Generally, linear plots over a broad pH range are obtained with hydraulic and ion-addition mobilization, whereas one-step CIEF in the presence of EOF generates linear plots over a very short pH range. It is noteworthy that hydraulic mobilization usually requires the use of sample additives to increase the viscosity of the sample matrix. The use of additives in the sample introduces a potential extra variable into the system, due to differences in preparation and storage of solutions, and disparities in batch-to-batch polymer production. However, for most applications, the effect of polymer differences is negligible, and the use of internal standards further corrects any variations.

Optimization of Chemical Mobilization

1.   Effect of Salt Type Used for Mobilization.   The effect of salt type in the overall resolution of hemoglobin variants during ion-addition mobilization is depicted in Figure 22 [307]. It is evident that mobilization is affected by the type of salt used to disrupt the chemical equilibrium attained during focusing. Sodium chloride was the first salt used for mobilization, and is still used in ion-addition mobilization, but it is clear that other salts provide better results; in particular, the use of

Figure 22   Effect of salt type used for mobilization. Human hemoglobin variants A, F, S, and C were focused for 4 min at 15 kV. Ion addition mobilization at 15 kV was carried out using different salts. (A) Mobilization using 100 mM NaCl. (B) Mobilization using 100 mM Na-acetate. (C) Mobilization using 20 mM Na-tetraborate. A 24 cm $\times$ 50 μm coated capillary was used. Detection at 280 nm.

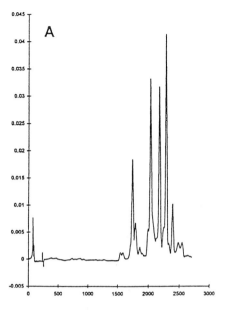

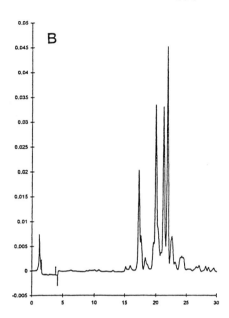

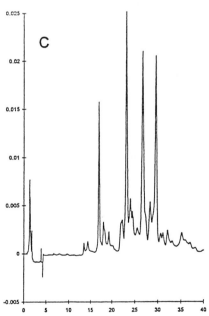

sodium tetraborate (Figure 22C) greatly increased resolution. Other salts produced modest improvements, and in some cases resolution was diminished.

2. Effect of Salt Concentration Used for Mobilization. The effect of the concentration of the salt used for mobilization is shown in Figure 23. Resolution was better at lower concentrations (especially at 25 mM) of mobilization salt, at the expense of total analysis time, but it is obviously an alternative when high resolution is essential.

Optimization of Hydraulic Mobilization

1. Pressure. Figure 24 [307] shows resolution as a function of the pressure applied during mobilization. Resolution clearly decreases as the pressure is increased. At 1 psi the sample components were detected as a group of fused peaks (panel B), and minor bands were totally lost. When the pressure is lowered to 0.5 psi the resolution was improved (panel A). If the same analysis was performed at a lower polymer concentration, the separation quickly diminished, thus, higher concentration of polymer (increased viscosity) counters the negative effects of higher applied pressure. But Figure 24 also shows that increasing polymer concentration could not totally restore the resolution to the level obtained when mobilizing using 0.5 psi, even when the MC was increased to 0.6%. At this polymer concentration mixing the sample and ampholytes was difficult, and the risk of entrapped bubbles increased significantly.

Minarik et al. developed a theoretical model [328] for pressure mobilization in uncoated capillaries. According to this model, increased peak broadening results when higher pressure and/or lower polymer concentration are used as a result of increased protein adsorption. To measure sample zone deformation due to the parabolic profile of laminar flow, mobilization was performed using pressure with the voltage turned off. Since peak broadening also occurs in coated capillaries (which reduce adsorption) the effect must be due to a combination of factors. Interestingly, the authors found that higher efficiency was obtained when pressure was applied to the

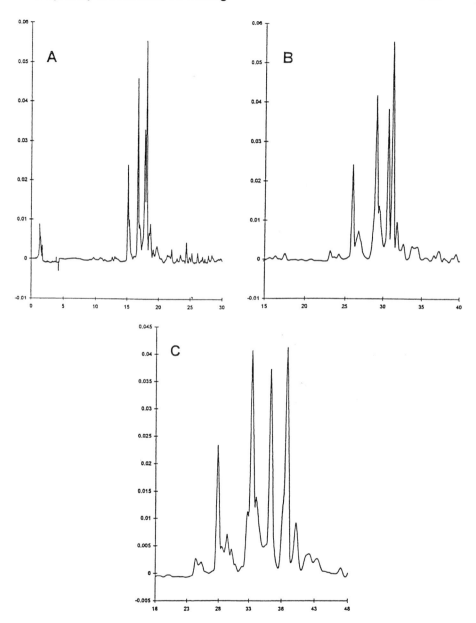

Figure 23 Effect of salt concentration used for mobilization. The mobilizer was prepared by adding the following concentrations of dibasic Na-phosphate to 40 mM NaOH: (A) 100 mM; (B) 50 mM; (C) 25 mM. All other parameters as described in Figure 22.

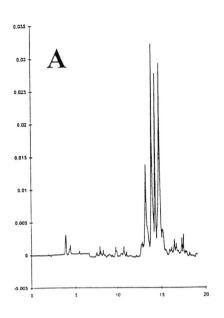

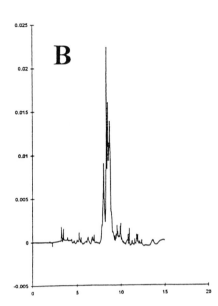

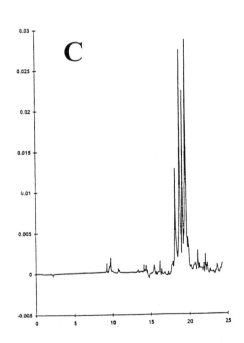

cathodic end of the capillary. The effect is attributed to interacting proteins becoming negatively charged, and thus being
electrostatically repelled from the capillary wall (which under those conditions is also negatively charged). As already
discussed, dispersion is reduced by maintaining the voltage
on during mobilization.

2. Gravity Mobilization. As stated above, resolution during
hydrodynamic mobilization is inversely related to the pressure applied. Gravity mobilization (Figure 11) as a means to
mobilize focused protein zones has been suggested [2] and
demonstrated previously [295]. Advantages of this method
include its simplicity (no valves, no pressure sources, no regu-
lators), and its universality (it can be implemented in almost
any CE instrument). The force applied can easily be controlled
by adjusting the height difference of the two reservoirs. Flow
velocity can be manipulated by adding viscosity enhancers
such as MC. When the method is optimized, high resolution
is achieved (Figure 25).

Effect of Viscosity

Viscosity differences within the capillary due to variations in protein
concentration or to sample composition may cause shifts in migration times, although the relative peak positions remain the same. For
this reason, the use of internal standards is strongly recommended,
particularly when the technique is used to estimate the isoelectric
point of an unknown protein. Since a more concentrated zone of

---

Figure 24 The effect of pressure in the performance of hydraulic mobilization. Human hemoglobin variants A, F, S, and C were mixed with
ampholytes (0.2% solids Bio-Lytes 3–10) containing 0.3% methylcellulose (4 K cps at 2%). After focusing for 4 min at 15 kV, the proteins were
mobilized applying pressure at (A) 0.5 psi, (B) 1 psi. The voltage was maintained at 15 kV during mobilization. An increased concentration of methylcellulose (0.6%) did not totally counter the adverse effect of higher (1
psi) applied pressure (C). Capillary: 24 cm × 50 μm, coated. Temperature
was thermostated to 25°C.

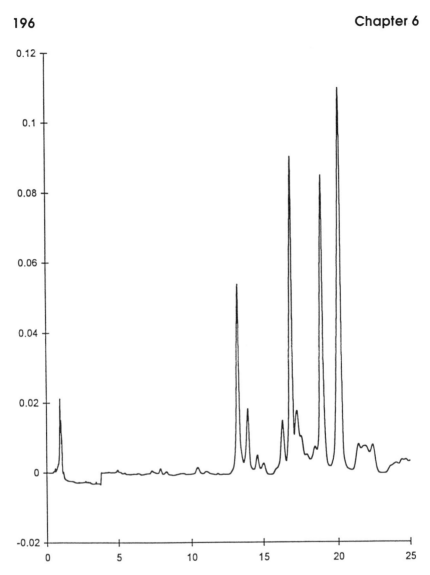

Figure 25 Analysis of hemoglobin variants by CIEF using gravity mobilization. After focusing, the catholyte (40 mM NaOH) vial was replaced with a vial containing a smaller volume of the same solution. Since the anolyte (20 mM phosphoric acid) vial was full, a height differential of about 2 cm was created. Conditions: 24 cm × 75 μm, coated capillary; focusing was carried out for 4 minutes at 15 kV. The sample was prepared in 2% solids Bio-Lytes 3–10 + 0.2% methylcellulose (4 K cps at 2%). The voltage was maintained at 15 kV during mobilization. Detection at 280 nm.

protein possesses a higher buffer capacity, chemical mobilization efficiency should also, theoretically, be affected. We have analyzed the same sample at concentrations varying from 5 to 500 µg/ml and did not observe any significant change in migration time (data not published). If a sample viscosity issues are suspected, adding hydrophilic polymers to the sample matrix usually corrects the problem. Care should be taken to maintain the same viscosity from sample to sample, and from day-to-day preparation of reagents. It should also be noted that focusing is delayed as the viscosity increases. Small amounts of polymers also may improve resolution by eliminating convection at the interface of the highly concentrated sample and the low-viscosity ampholyte solution, and by reducing diffusion. When CIEF is performed in uncoated capillaries, or in any capillary that exhibits significant EOF, adding hydrophilic polymers to the sample + ampholyte matrix is a routine practice used to manipulate the speed of EOF. Viscosity has a significant impact in CIEF with hydraulic mobilization, and a useful rule of thumb is that resolution increases with increased viscosity, at the expense of analysis time. Viscosity variations are also tightly related to temperature variations.

Effect of Variable EOF

Variable electroosmotic flow can contribute to poor reproducibility in CIEF. When using coated capillaries, this may be caused by loss of capillary coating after prolonged use, particularly at the cathodic end where the coating is continually exposed to high pH conditions. Increased EOF can be recognized by the delayed appearance of peaks during the focusing step, and early migration of peaks during mobilization, accompanied by loss of resolution. Well-coated capillaries should have lifetimes of up to several hundred analyses, depending on analysis conditions. Poor separation due to protein adsorption or precipitation may be misdiagnosed as coating failure; thus, after extensive washing, the coating performance can be evaluated separately through the use of neutral markers (e.g., niacinamide at pH 8.5) under free zone electrophoresis conditions to measure the velocity of electroosmotic flow.

Gradual adsorption of protein to the wall of polyacrylamide-coated capillaries that causes deterioration of capillary performance in CIEF can be minimized by purging the capillary between each analysis with one or more wash reagents, such as dilute acid (e.g., 10 mM H_3PO_4) or surfactants (e.g., 1% Brij). A capillary with degraded performance can often be regenerated with extended washes with these same solutions. In cases of serious performance deterioration, a brief wash with 1% SDS in basic solution (NaOH, pH 10) or washing with 0.1% trifluoroacetic acid in 40:60 water/acetonitrile may regenerate the column. The capillary should be washed extensively with water following exposure to aggressive wash solvents. Best lifetime for coated capillaries is achieved by washing the capillary with dilute acid and water after usage, followed by purging with dry nitrogen for 5 min before storage. For uncoated capillaries, protein adsorption and viscosity changes of the solutions used results in EOF irreproducibility, and thus, variable migration times. Uncoated capillaries can be washed with harsh solutions (e.g., 100 mM NaOH) to restore the original EOF velocity.

Effect of Applied Voltage

Theoretically, best resolution is obtained at high voltages. But in practice, variation of the electric field intensity under typical analysis conditions for CIEF (300–1000 V/cm) is a parameter of relatively minor importance in optimizing resolution. Field strength should be kept low enough to avoid excessive Joule heating, particularly at the beginning of focusing. If reduction in analysis time is desired, high field strengths will shorten focusing time but will effect mobilization time only when chemical mobilization or single-step CIEF in the presence of EOF is used.

If the initial current is high, and Joule heat problems arise, CIEF can be performed at a constant current instead of constant voltage, setting a higher voltage limit at a desired value. Under these conditions, as the current decreases the voltage increases, until it reaches the set limit. At that point the voltage remains constant and the current continues decreasing. Initial high current during focusing is a sign of high salt content, and protein precipitation and current breakdown may occur.

Effect of Capillary Temperature

Protein solubility is affected by the temperature of the solution. As described above, during the CIEF process proteins become highly concentrated at their isoelectric point, with low concentration of salt ions, and therefore they may tend to precipitate. The temperature of the capillary can be manipulated to increase their solubility, but it should no be too high as protein denaturation may occur. Higher temperature is used mainly in combination with other solubilizing agents, and temperature may be used to increase the solubility of these additives (e.g., urea) rather than the solubility of the proteins.

Since pI, mobility, and viscosity are all affected by temperature, once again, the use of internal standards is recommended. However, the use of synthetic pI standards to estimate pIs should be approached with caution since the pIs may not be affected equally by temperature. Since temperature has a direct effect on viscosity, all effects of viscosity on the CIEF process apply to temperature changes.

The Use of Internal Standards

The development and use of CIEF as a routine analysis technique has been slow, mainly because its run-to-run and inter-day reproducibility is usually not as good as other modes of CE. The degree of variation is especially acute with migration times. We have already pointed out some sources of variability which can be minimized by using internal standards. The use of standards allows results to be normalized, but more importantly, it allows protein bands to be identified by their pI value, instead of using the sometimes unpredictable migration time. An important aspect of reporting compounds by their isoelectric point is that it does not depend on analysis conditions (although it is affected by ampholyte composition), and variations due to changes in EOF, viscosity, capillary length or internal diameter, etc. are readily corrected.

Important characteristics of internal standards include high purity, stability, availability of species with known pI values spanning the pH range of interest, high absorption at the detection wavelength, and nonreactivity with sample components and/or ampholytes. Protein standards are widely used in slab gel IEF, and they are available from multiple commercial sources. One drawback of protein standards is that they are only available as premixed solutions of a fairly high number of proteins, and they are intended to be applied in a single lane of a slab gel. For CIEF, it is preferable to combine the sample and standard, but the complexity of the resulting pattern often makes identification of the compounds of interest very difficult, if not impossible. Single proteins usually lack the necessary purity, and most of them present several bands (including protein and low-molecular-weight contaminants), complicating the separation pattern.

A family of substituted aromatic aminophenol compounds has been synthesized [329], and they possess all of the desirable characteristics of internal standards. Since these compounds have very high UV absorption at 280 nm, they can be introduced into the capillary as a second injection (occupying only a small portion of the capillary), allowing recovery any unused portion of the sample without contamination. Furthermore, they are small molecules that can be removed easily by dialysis (an important factor to consider when collecting fractions). Figure 26 displays a CIEF analysis of human hemoglobins A and S flanked by synthetic 5.3 and 8.4 pI markers. Table 2 shows the reproducibility (%RSD) obtained by analyzing the

Table 2A Migration Time Reproducibility

Run	Hb A	Hb S
1	19.16	18.77
2	19.01	18.63
3	18.88	18.50
4	18.74	18.36
5	18.63	18.26
Average	18.88	18.50
Std. Dev.	0.21	0.20
%RSD	1.11	1.10

Table 2B Isoelectric Point Reproducibility

Run	Hb A	Hb S
1	7.39	7.22
2	7.38	7.23
3	7.38	7.22
4	7.38	7.22
5	7.38	7.22
Average	7.38	7.22
Std. Dev.	0.004	0.004
%RSD	0.06	0.06

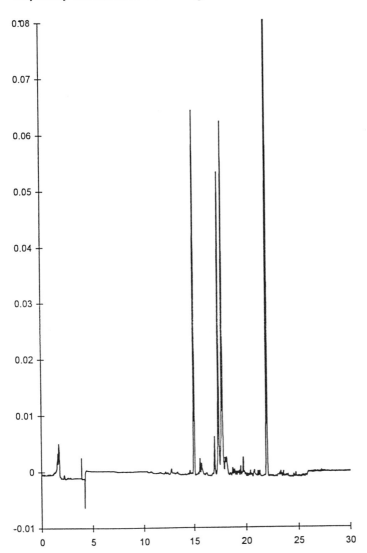

Figure 26 Human hemoglobins A and S flanked by synthetic markers of pI 5.3 and 8.4. After the capillary was filled with sample dissolved in ampholytes, the standards (also mixed with ampholytes) were introduced into the capillary as a second, short injection. With this procedure, different markers can be used with the same sample solution. A better estimation of protein pI is achieved if the pI markers are as close as possible to the sample of interest. Analysis conditions similar to those of Figure 18 except that the capillary used was 24 cm × 50 μm.

migration times, and by using the pI standards to calibrate the protein zones [307,322]. The %RSD values reported for pI were extremely good (0.061%) as compared with migration times %RSD (1.1%). Software packages that identify the internal standards, plot their migration time as a standard curve and automatically report the pI of "unknowns" will make the CIEF process simpler, more powerful and reliable.

The Use of Reagents with Buffering Capacity

Reproducibility is also affected by continuous use of reagents lacking buffering capacity. In one model for IEF [282], the anolyte and catholyte solutions provide a "pH cage" from which zwitterions (including ampholytes) cannot escape either by electromigration or diffusion (see Figure 1). This cage can contain only molecules with lower (higher) pI than the catholyte (anolyte). If the pH of the catholyte (anolyte) decreases (increases) all molecules with a higher (lower) pI will migrate out of the capillary toward the electrode. Sodium hydroxide is widely used in both gel IEF and CIEF, but solutions of this compound lack pH buffering capacity. Using a simple system (hemoglobin variants) a systematic drift of migration time as a function of time can be observed (Figure 27). Migration time (MT) drift is accompanied by loss of resolution (Figure 28), but MT can be restored almost to its original value (see last data point of Figure 27), and resolution improved by simply replacing the catholyte with a fresh solution. Capillary IEF could certainly benefit if buffering solutions can be employed. One difficulty in the use of compounds other than strong bases as catholytes is that most of them are salts. Although in theory a pH cage can be formed between any two buffers of different pH, the use of most buffers interferes with the focusing process, and in many cases it is impossible to achieve equilibrium. It is well known that most salts can actually be used to disrupt the focusing equilibrium by the process termed ion-addition mobilization (see the earler section on mobilization).

Progressive flow of nonproton cations (anodic mobilization) or nonhydroxyl anions (cathodic mobilization) causes a progressive pH shift along the capillary, resulting in mobilization of proteins in sequence past the detector point. So any buffer solution containing a competing ion will cause mobilization (or drift). On the other hand,

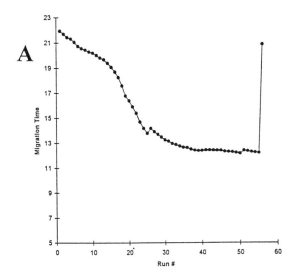

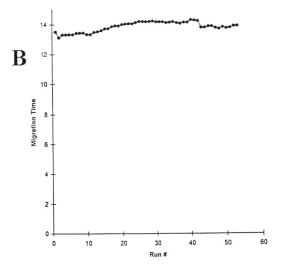

Figure 27 Migration time drift after prolonged use of (A) NaOH catholyte, and (B) taurine buffer catholyte. Plots of migration time vs. run number were generated after analyzing bovine hemoglobin continuously for over 50 runs using the same anolyte, catholyte, and mobilizer. Focusing: 4 min at 15 kV. Capillary 24 cm × 50 μm, coated. The sample was prepared in 2% solids Bio-Lytes 3–10. The catholyte was changed for a fresh solution in the last run.

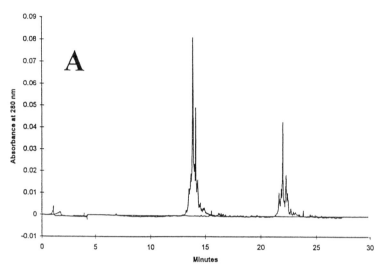

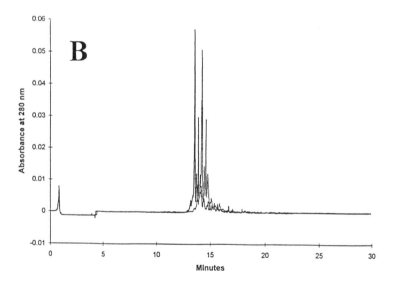

Figure 28 (A) Electropherograms depicting runs 1 and 25 using NaOH as catholyte. Migration time drift is accompanied by loss of resolution. (B) Electropherograms depicting runs 1 and 25 using taurine as catholyte. The protein pattern obtained is remarkably maintained even after prolonged use of the buffered catholyte. All other conditions as described in Figure 26.

one way to maintain electroneutrality during CIEF would be to add a neutral (zwitterionic) compound to any of the sides of the equation (as long as the added compound is neutral at the pH of the solution to which it is added)

$$C_{H}^2 + \Sigma C_{NH_3^+} = C_{OH^-} + \Sigma C_{COO^-} + C_{xn^+ym^-} \qquad (6\text{-}5)$$

where $C_{xn^+ym^-}$ represents an electrically neutral molecule, and if this compound possesses buffering capacity it could eliminate or reduce MT drift. Notice that a neutral compound could be added to either side of the equation without disrupting electroneutrality. If this ideal situation cannot be implemented, other compounds can be used as buffers even if they are not neutral, as long as they allow the focusing of the proteins to proceed "normally." To date, the buffers we have used with success exhibit relatively low mobility. Figure 27 displays MT over 50 consecutive analyses using $0.5\,M$ taurine buffer pH 8.8 as catholyte solution [307]. The same vial (0.5 ml) was used for all analyses. Although there is still some variability, the general downward trend was eliminated, and resolution was not affected over time (Figure 28B). Although taurine is not neutral at pH 8.8, it still allows the focusing of the sample. The concept of using neutral molecules as catholyte in conventional electrophoresis has been explored in the past. In most cases the catholyte consisted of amino acids (e.g., arginine, lysine) or related molecules (e.g., lysyl-lysine). Using the same concept, we have been able to use buffers other than taurine, including MES, CAPS, acetate, MOPS, etc. This approach was described previously by Nguyen and Chrambach [330] for gel electrophoresis. Besides the advantage of higher reproducibility, the use of buffers also minimizes the deleterious exposure of coated capillaries to the harsh conditions of extreme pH typical of NaOH solutions. An interesting feature of the catholyte buffers so far used in our lab is that the pH can be varied within at least a small range (we have used taurine buffers ranging in pH 8.5–8.9) without noticeable adverse effects. This technique can potentially improve results when single-step CIEF is used, since in this technique the catholyte is used to fill a significant portion of the capillary, and the column contents are transported into the outlet reagent (typically the catholyte) during the analysis.

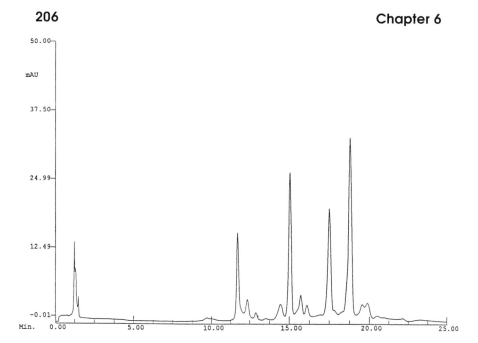

Figure 29 Nonequilibrium single-step CIEF of hemoglobin variants A, F, S, and C using taurine buffer. Buffers, such as taurine, which are not isoelectric, interfere with the attainment of equilibrium during focusing. If the analysis is prolonged, the focused zones migrate pass the detector point. This example of nonequilibrium IEF works best for neutral and basic proteins. Conditions: 22 cm × 50 μm, coated capillary; detection at 280 nm; 20 mM phosphoric acid was used as anolyte; temperature set at 25°C. Analysis was carried out at 15 kV using 40 mM taurine buffer, pH 8.8 as catholyte.

When nonzwitterionic compounds are used, true equilibrium is not reached since focused bands eventually migrate toward the detector if the voltage is maintained (Figure 29). Thus, this method is a variation of single-step CIEF (or ITP). When the migration of proteins is too slow (e.g., for acidic proteins), the transport can be accelerated by switching to mobilization after focusing has been achieved.

CAPILLARY IEF AS A MICROPREPARATIVE TECHNIQUE

Ever since capillary electrophoresis was introduced, the desire to collect fractions of pure sample components has existed. The extremely small quantities injected into the capillary in CZE has limited the number of applications where sufficient amount can be recovered. Capillary IEF not only utilizes the whole length of the capillary for injection, but protein zones are highly concentrated during the analysis. Thus, CIEF has potential as a micropreparative technique, especially for enzymes, since only very small quantities of these polypeptides are required for enzymatic reactions. It is noteworthy that denaturation may occur when collecting fractions into NaOH (or any other extreme pH solution). Buffered catholytes should be used to avoid protein denaturation.

Capillary IEF was used to separate and collect fractions of model proteins from a single run [331]. To reduce the velocity of the EOF in the uncoated capillary (88 cm × 100 μm), 0.4% HPMC was added to all solutions. After focusing at 30 kV for 15 min, pressure mobilization (50 mbar) was used to transport the focused zones to the fraction collector, which contained carrier ampholytes (15 μl of 1.25% Servalyte per vial). The collected proteins were then subjected to amino acid sequencing for up to 45 cycles.

APPLICATIONS

A variety of CIEF applications has been published, including the analysis of human transferrin isoforms [289], recombinant proteins (e.g., human recombinant tissue plasminogen activator [332]), human growth hormone [333] (Figure 30), γ-globulins [333], and hemoglobin variants [270,271]. As indicated by these reports, CIEF analyses are used to characterize proteins, as well as to determine their purity. Also, it has been suggested that conventional IEF in gels can distinguish conformational states of proteins [334], although the same has not yet been reported for CIEF. Capillary IEF is powerful tool for the detection of protein modifications, such as deamidation, deletions, insertions, proteolytic clips, N- or C-terminal modifications, and glycosylation [286].

Hemoglobin Analysis

Hemoglobin Variants

Analysis of hemoglobin (Hb) variants is of major importance in clinical diagnostics. Hemoglobin is a tetramer of four globin chains: two

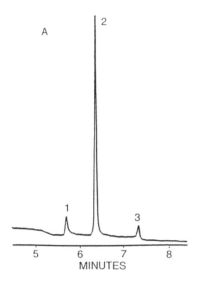

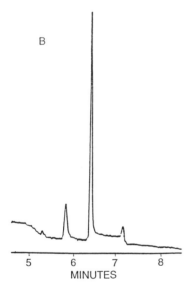

Figure 30 Capillary IEF of human growth hormone. (A) Moderate aging. Peak 1: deamidated hGH; peak 2: native hGH; peak 3: hGH fragment. (B) hGH after prolonged aging. Analysis conditions: 12 cm × 25 μm LPA coated capillary; detection at 280 nm; focusing at 8 kV for 5 min; cathodic mobilization by ion-addition, at an applied field of 8 kV. (Reproduced from Ref. 333 with permission.)

α-globin, and two β-, δ-, or γ-globin chains. Capillary IEF can easily resolve adult Hb (Hb A) from normal variants such as fetal Hb (Hb F) which is found in blood up to the age of six months after birth at which time it is replaced by Hb A. Capillary IEF can also be used to distinguish abnormal Hb species associated with a variety of blood disorders. Disorders arise from defective genes that code for altered Hb chain sequences, and abnormal levels of individual globin chains produce novel hemoglobin tetramers characteristic of the disorder. A high number of point mutations (about 600) in the Hb chains has been estimated [335].

Hemoglobin is very soluble in water, and is present in erythrocytes at a concentration that can exceed 300 mg/ml; therefore, it is not prone to precipitate under CIEF conditions. Each of the hemoglobin variants usually vary only slightly in composition and are difficult to separate with other modes of CE. Capillary IEF can resolve proteins with pIs that differ by as little as 0.02 pH unit. This power of resolution is more than enough to detect subtle shifts in pI in the major Hb variants present in a sample. Figure 31 shows an excellent separation of hemoglobins A, F, S, and E with isoelectric points of 7.10, 7.15, 7.25, and 7.50, respectively.

α-Thalassemias

Abnormalities in the production of globin chains can produce blood disorders collectively known as thalassemias. Thalassemias usually arise from the deletion of one or more of the four globin genes. α-Thalassemias affect the production of α-globin chains, and the resulting disorder can range from asymptomatic to lethal anemia. Capillary IEF allows for rapid analysis of hemoglobin (Hb) disorders such as Bart's disease, in which the tetramer is composed of four γ-globin chains, and Hb H which is characterized by a tetramer of four β-globin chains. Figure 32 displays a CIEF analysis of an Hb Bart's patient sample [274]. Individual globin chain analysis can also be carried out using capillary zone elctrophoresis [270].

In a similar application with a different approach [336], human Hb variants (A, A1, A1c, A2, F, S, D, and E) were analyzed by CIEF with electroosmotic transport. For different samples examined, good correlation between CIEF, slab gel IEF, and HPLC was observed. HPLC runs were found to be time consuming as compared with CE. Since a higher number of samples can be analyzed in a gel, this method

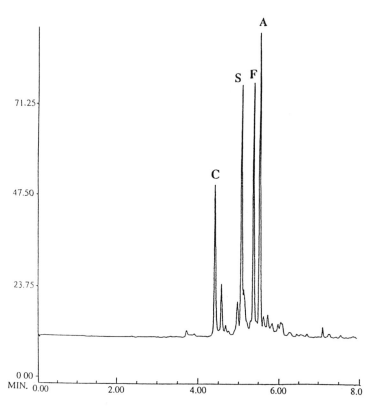

Figure 31 Capillary IEF of Hemoglobin A, F, S, and E reference standard. Analysis performed in a 24 cm × 50 μm capillary internally coated with polyAAEE. Liquid capillary cooling at 20°C. Hemoglobin samples were mixed with pH 3–10 ampholytes (Bio-Lytes 2%). Focusing at 10 kV for 5 min. Ion-addition cathodic mobilization at 10 kV. On-line detection at 415 nm.

was found to be the one with higher throughput, but it did not provide quantitative information. Important factors that needed optimization in CIEF were carrier ampholyte selection and concentration. Best resolution was obtained using a 1:2 v/v mixture of broad range (3.5–10) with narrow range (6.7–7.7) ampholytes.

Hemoglobin variants have been quantified by CIEF with hydraulic mobilization by Hempe and Craver [337]. The aim of this study

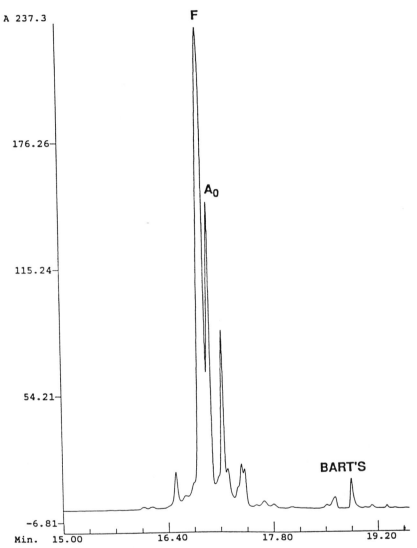

Figure 32 Capillary IEF of hemoglobins from a patient with hemoglobin Bart's disease. Analysis performed in a 17 cm × 25 μm capillary internally coated with linear polyacrylamide. Liquid capillary cooling at 20°C. Hemoglobin samples were mixed with pH 3–10 ampholytes (Bio-Lytes 2%). Focusing at 10 kV for 5 min. Ion-addition cathodic mobilization at 10 kV. On-line detection at 415 nm. (Reproduced from Ref. 274 with permission.)

was to validate CIEF as a routine technique for clinical analysis, including quantitation of human hemoglobin variants. Hemoglobin variants present in patient samples were identified by linear regression of pI vs. migration time using Hb A as a reference peak and external calibration with Hb standards. Quantitative CIEF results for Hb A2, S, F, and A correlated well with results obtained by other methods used by a reference laboratory. Resolution was demonstrated by separation of Hb S and Hb D-Los Angeles (calculated pIs 7.207 and 7.181, respectively). The authors suggested that CIEF could replace multiple conventional assays, simplifying both the operative and regulatory aspects of the analysis. The quantitative advantages of CIEF over slab gel IEF in clinical diagnostics were also demonstrated by these authors [337]. Clinical diagnostics requires increasingly powerful and specific separation techniques but these methods must be highly reproducible. The analysis of hemoglobins by CIEF using standards to identify variants based on their pI values consistently produced coefficients of variation (CV) less than 0.05%. Neonate blood normally contains 60–90% hemoglobin F, and about 10% is present as Hb Fac. In most techniques used for hemoglobin analysis, Hb Fac comigrates with Hb A (which in many cases is at a similar concentration of Hb Fac), thus preventing the determination of heterozygous A/S and homozygous S/S. Since CIEF could resolve all these Hb variants, this technique can potentially be used to screen for neonatal sickle cell anemia hemoglobinopathy. Another application for clinical diagnostics was the determination of Hb A_{1c}, which is widely used for long-term glycemic control in diabetic patients. Hemoglobin quantification is a key parameter in this type of analyses, and the technique must allow accurate estimation of components present in small amounts. Capillary IEF allowed quantitation of components present in concentrations below 0.5% of total hemoglobin, with CVs between 2 and 11%.

Capillary IEF was also used to follow the process of blood replacement in calves [338]. Young bovines were studied as potential blood donors for other bovines used for artificial organ research. Collection of blood from the studied bovines was shown to alter the patterns of hemoglobin production (with a decrease in concentration of Hb A and the appearance of several unidentified peaks), which were restored to normal after about 18 days.

Glycated Hemoglobins

The potential of CIEF as a clinical diagnostic tool for the determination of Hb A_{1c} as reported by Hempe and Craver [337] was described above. To improve resolution, Conti et al. [339] decreased the slope of the pH gradient in the region around pH 7 in order to resolve glycated hemoglobins by incorporating 0.33M 6-aminocaproic acid and 0.33M β-alanine in a mixture of ampholytes (Ampholine 6-8, and Pharmalyte 3-10). The system also contained short chains of linear polyacrylamide, and 50 mM lysine was used as catholyte to improve reproducibility and increase capillary lifetime. In the absence of β-alanine and 6-aminocaproic acid, CIEF was unable to resolve HB A_{1c} in blood of a known positive diabetic patient (Figure 33). CIEF analyses were performed in capillaries internally coated with N-acryloylaminopropanol (AAP). UV absorption at 416 nm was employed for detection. The preparation of hemoglobin from red blood cells consisted of washing the cells three times with isotonic saline solution; lysis was then accomplished by placing the cells in a 5% KCl solution. After centrifugation at 18,000 g for 12 min, the supernatant was gassed with CO for 1 min to prevent oxidation of the Hbs. Samples for optimization of the CIEF method were obtained by small scale fractionation in immobilized pH gradients (pH range 6.8–7.8). Best results were obtained when the focused zones were mobilized using the catholyte + 20 mM Na-Phosphate. The CIEF estimated values for HB A_{1c} were then compared to Helena's gel electrophoresis system, which is used as a standard method for quantitation of HB A_{1c}. Both techniques produced similar values for 15 different samples analyzed. The significance of this agreement in values is magnified because this was a blind test. Mobilization using NaCl quickly deteriorated the capillary's internal coating, and also produced more oxidized species of Hb. The high reproducibility obtained was due in part to the mode of injection (filling up the whole capillary instead of loading a sample plug), the quality of the internal coating (the authors used capillaries with EOF below 9.7×10^{-7}

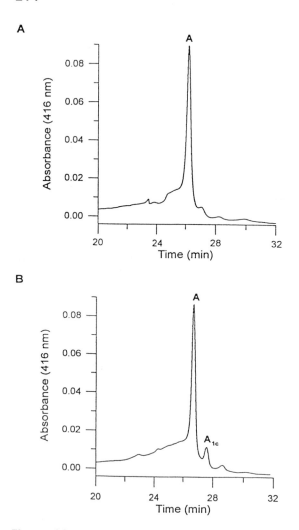

Figure 33 Effect of "separators" in the performance of CIEF. The sample, obtained from diabetic patient, was known to contain Hb A_{1c}. Panel A: Capillary IEF was performed using a mixture of 5% Ampholine 608, 0.5% Pharmalyte 310, and 3% short-chain polyacrylamide. Panel B: 0.33 M β-alanine and 0.33 M 6-aminocaproic acid were added to the ampholyte mixture to flatten the pH gradient. Note that when the separators were added, the peak shape and resolution of HbA_{1c} was improved. Anolyte: 50 mM acetic acid, pH 3.5; catholyte: 50 mM lysine, pH 9.7. Focusing at 10 kV for 5 min. Mobilization by a combination of ion-addition (catholyte + 20 mM phosphate) and siphoning. (Reproduced from Ref. 339 with permission.)

$cm^2 V^{-1} s^{-1}$.), and to the elimination of nonbuffering ions from the CIEF system (including the elimination of TEMED as spacer, and NaCl as mobilizer).

Haptoglobin–Hemoglobin Complexes

If no mobilization step is initiated, the focused zones remain in a fixed position in the pH gradient for a relatively long period of time. Righetti et al. [340] exploited this characteristic of IEF to monitor the complex formation of hemoglobin and haptoglobin. Haptoglobin is a serum glycoprotein which exhibits strong affinity for hemoglobin and appears to participate in the cycling of heme iron by forming an irreversible noncovalent complex with hemoglobin released into the plasma by lysis of red cells. The complex is readily absorbed into the liver, where it is catabolized, effectively conserving iron ions that otherwise would be excreted. The extreme affinity of haptoglobin is directed to the α-chain of the globin portions of hemoglobins A, F, S, and C. Therefore, haptoglobins do not bind to methemoglobin, heme, or unusual forms of hemoglobins in which the α-chain is missing.

In a 24 cm × 75 µm capillary coated with poly-*N*-acryloylaminopropanol, a known amount of hemoglobin was focused for 10 min at 10 kV in a 6–8 pH gradient (5% pH 6–8 + 0.5% pH 3–10 Ampholines containing 3% short-chain polyacrylamide) using 50 mM lysine (pH 9.7) as catholyte and 50 mM acetic acid (pH 3.5) as anolyte. Isoelectric lysine was used to increase the buffering capacity of the catholyte. After focusing, increasing amounts of haptoglobin were electrophoretically introduced into the capillary from the cathodic end. When haptoglobin reached the focused hemoglobin, it formed a complex with an approximate pI of 5.5. Since the pI of the complex was outside the pH range of the ampholytes, it exited the capillary toward the anode, which was located at the detector (UV absorption at 416 nm) end of the capillary. Thus the method could be used to quantify the amount of haptoglobin in human sera in pathological conditions such as hemolytic anemia and transfusion reactions. Residual focused hemoglobin was mobilized by adding 20 mM NaCl to the anolyte, coupled with siphoning (650 µl in the inlet, 450 µl in the outlet) while maintaining 10 kV. The capillary was thermostated to 15°C, while the samples and reagents were maintained at 6°C.

Immunoglobulins

Immunoglobulins in the form of monoclonal antibodies are manufactured commercially for therapeutic and diagnostic uses. Major areas of consideration are, thus, quality control and bioactivity. Separation of immunoglobulins has proven to be a challenge, even for well-established techniques like HPLC. Difficulties arise due to the large size of the antibodies, and to their surface properties, which increase their tendency to interact. Monoclonal antibodies have been shown to possess microheterogeneity due to posttranslation modifications such as glycosylation. Gel IEF is used routinely to analyze different batches of antibodies, but this type of analysis presents several drawbacks already discussed. Figure 34 shows that CIEF can be used for the analysis of antibodies. One major difficulty is maintaining the antibodies in solution; the use of additives, short focusing time, low sample concentration, and other precautions (see Optimizing CIEF Separations) are a major part of method development to achieve reproducible high-resolution separations. Capillary IEF of monoclonal antibodies with pIs near neutral pH was carried out in single-step mode in the presence of EOF, and using protein markers as internal standards [341].

Capillary IEF was also used to follow the production of recombinant antithrombin III (r-AT III) by cultures of hamster kidney cells [342]. Recombinant-AT III inhibits serine proteases such as factors IXa, Xa, and XIa and thrombin. Thrombin is the main target of r-AT III, and it was found in this report that heparin (a polysaccharide) of at least 18 units in length enhances the rate of inactivation. Interference by the media from which the samples were collected posed some difficulties since some of the media components have similar characteristics to those of the compounds of interest. Capillary IEF was used to determine the pIs of the separated components after sample purification by HPLC. Three major bands showed pIs of 4.7, 4.75, and 4.85, and three minor peaks had pIs of 5.0, 5.1, and 5.3. These data closely resembled data already published for serum AT III based on conventional IEF.

The feasibility of using CIEF for analysis of monoclonal antibodies in a quality control environment was demonstrated for recombinant humanized monoclonal antibody HER2 (rhuMAbHER2) by Hunt

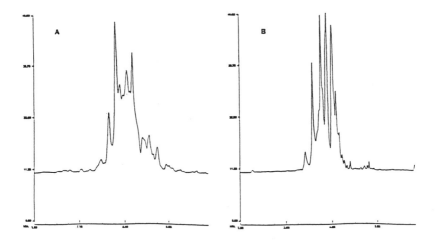

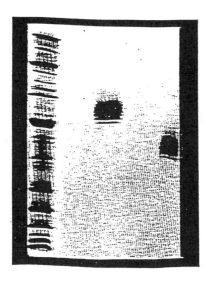

Figure 34 Capillary IEF and slab gel IEF analysis of antibodies. Two murine IgG monoclonal antibody preparations were analyzed first by gel IEF. Clusters of 6–8 bands were obtained after silver staining. Analysis by CIEF revealed a similar pattern of 6–8 major bands plus several minor components not observed in the gel. Analysis conditions identical to those used in Figure 30. (Reproduced from Ref. 333 with permission.)

et al. [343]. This protein is present in increased concentration in certain cancers (e.g., breast cancer). Besides primary structure heterogeneity (combinations of 214-residue light chains and 449- or 450-residue heavy chains), rhuMabHER2 can exhibit charge differences due to deamidation or C-terminal clipping. Resolution of the five observed components was optimized by mixing Pharmalyte 8-10.5, Bio-Lyte 3-10, and Bio-Lyte 7-9 ampholytes in an 8:1:1 ratio. Figure 35 shows a very good correlation of the IEF and the CIEF pattern

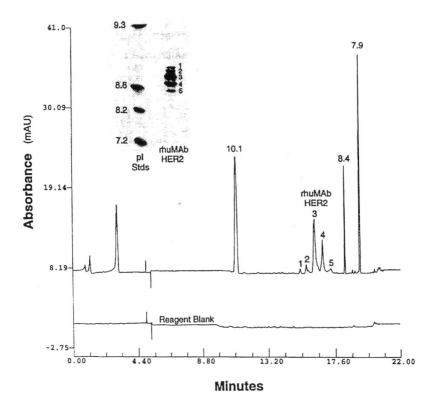

Figure 35 Gel IEF and capillary IEF of rhuMAbHER2. The top trace of the electropherogram depicts rhuMAbHER2 flanked by pI markers. The bottom trace shows a blank of the same sample. Note the excellent correlation of protein species resolved and their relative proportion for the gel and capillary IEF. Also, the pI of the major species, as determined by the extrapolation of migration times of rhuMAbHER2 and pI markers, are in agreement. (Reproduced from Ref. 343 with permission.)

obtained for rhuMAbHER2. The isoelectric points of the major bands were determined through the use of internal standards, and the values obtained correlated well with the values obtained from gel IEF. The method was capable of revealing differences due to storage conditions at 5 and 37°C (Figure 36), and the changes observed (increased acidic peaks) were consistent with protein deamidation. Intra-assay reproducibility ranged from 0.7 to 0.9% RSD for migration time, 0.8 to 3% RSD for peak area, and from 1 to 3.7% for area percent. Inter-assay reproducibility for migration time varied from 0.4 to 0.6% RSD, 1.2 to 3.2% for peak area, and 1.1 to 4.2% for area percent. All analyses were performed in the same coated capillary.

Due to the generally low concentration of contaminants, an important parameter for QC labs is the limit of detection. For rhuMabHER2 this limit was estimated to be 2 ppm. Care should be exercised when analyzing low-concentration samples, since ampholytes may show residual absorption even at 280 nm [308]. This problem can be reduced by decreasing the ampholyte concentration (e.g., to 0.5% w/v).

Glycoform Analysis

A glycoprotein may vary in the location, length, and composition of sugar moieties attached to the polypeptide chain. The saccharide component of these glycoforms may play important roles in cell recognition, protein function, stability, solubility, and immunogenicity. In the development and manufacture of recombinant protein therapeutics, the distribution of glycoforms can therefore determine the efficacy and stability of the product. An understanding of the sugar content of recombinant proteins is particularly important since the glycosylation pattern is defined by an organism different from the end beneficiary. Introduction of carbohydrate groups can produce subtle changes in the protein isoelectric point, so IEF is a standard method for characterization of glycoforms; CIEF provides an automated quantitative method for glycoform analysis. Other applications of glycoform analysis include determination of hemoglobin A1c to monitor diabetes miellitus and determination of elevated transferrin glycosylation as an indicator of alcoholism and pregnancy.

An example of two-step CIEF applied to the analysis of recombinant proteins is the fractionation of human recombinant tissue

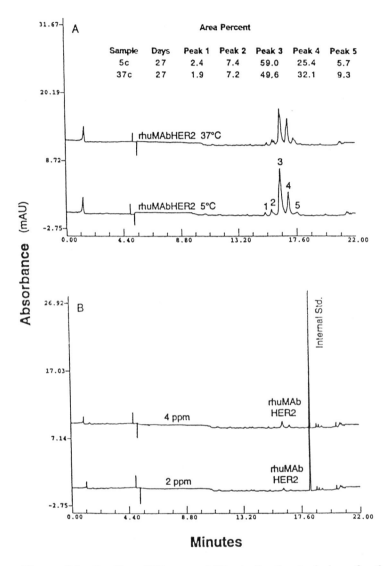

Figure 36 Capillary IEF as a stability-indicating technique for the analysis of rhuMAbHER2. Top panel: The sample was stored at 5°C or 37°C for 27 days. Although the number of species remained constant, the ratios of the peak areas changed, particularly the decreased amounts of peak 3 and the increased amounts of peak 4. Since peak 4 is more acidic, this shift was consistent with protein deamidation. Bottom panel: the limit of detection of this CIEF method was determined to be 2 ppm. (Reproduced from Ref. 343 with permission.)

plasminogen activator (rtPA) glycoforms [332]. Tissue plasminogen activator is a protein that degrades blood clots, and its recombinant form is produced for the treatment of myocardial infarction. This 59 kDa glycoprotein possesses three N-glycosylation sites. Type I rtPA is glycosylated at all three sites (residues 117, 184, and 448), whereas type II rtPA is glycosylated at two sites (117 and 448). Although rtPA was purified extensively to yield high purity of the polypeptide, in some instances up to 20 peaks were observed during CIEF (Figure 37). Treatment of rtPA with neuraminidase (an enzyme that removes sialic acid residues) greatly simplified the pattern, suggesting that heterogeneity is due to the variation of sialylation. CIEF performance was suitable for validation of the technique as a routine test [344].

CIEF analysis of rtPA in the presence of urea can also be carried out in an uncoated capillary using pressure mobilization [345]. The final urea concentration used was 4 M, and EOF was reduced by adding polymers to the reagents and sample (0.4% HPMC produced better results than PEG).

There is yet another CIEF variation for the analysis of rtPA [346]. The one-step CIEF method described by Moorhouse et al. was performed in a neutral coated capillary and through the addition of HPMC a constant residual EOF was achieved (Figure 38). The sample was prepared by dilution to 125–250 µg of protein per ml in 3% ampholytes 3–10 + 5–8 (50:50) containing 7.5% TEMED and 4 M urea. Results obtained by CIEF correlated well with those generated by IEF, and the analysis was completed in less than 10 min. The profile observed in Figure 38 suggests that in this analysis the peaks represent migration of proteins past the detection point from the detector-distal segment of the capillary during the focusing process. The decrease in baseline amplitude after migration of the protein peaks is characteristic of a focusing profile.

Transferrin and Other Metalloproteins

Metalloproteins contain one or more metal ions as part of the protein structure. The metal bound to the polypeptide chain usually acts as a catalytic or structural cofactor. These proteins participate in many cell processes, including metabolism, gene expression, and respiration. Examples of metalloproteins analyzed by CE include

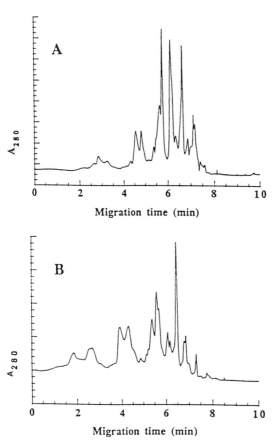

Figure 37 Microheterogeneity of human recombinant tissue plasminogen activator (rtPA) as analyzed by CIEF. Panel A: Type I rtPA; panel B:
Type II rtPA. Analysis performed in a 14 cm × 25 μm coated capillary
(Bio-Rad Laboratories); voltage: 12 kV focusing (2 min), 8 kV mobilization. Ampholyte solution: 2% total solids concentration (6–8) + 2% CHAPS
+ 6 M urea. Detection at 280 nm. (Reproduced from Ref. 332 with permission.)

hemoglobin, carbonic anhydrase, transferrin, conalbumin, myoglobin, albumin, alkaline phosphatase, metallothioneins, and ferritin.

Transferrin is an iron-transporting protein present in blood plasma.
This protein is capable of binding up to two ferric ions. Besides differences in iron content, transferrin molecules differ in the amount
and composition of sugar groups. Analysis of human transferrin

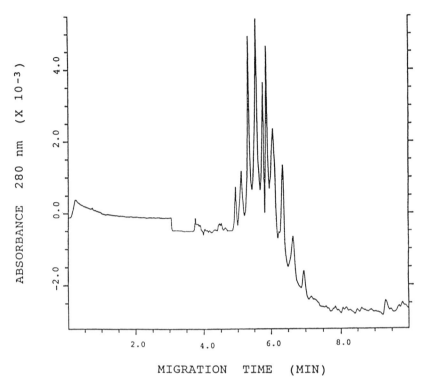

Figure 38 Single-step CIEF of rtPA. The analysis was performed in a 27 cm × 50 μm coated capillary, thermostatted at 20°C. The separation was achieved using reverse polarity (anode at the outlet vial, which contained 20 mM phosphoric acid) at 500 V/cm. Sodium hydroxide (20 mM) was used as catholyte. The rtPA sample was diluted to 125–250 μg/ml in a solution containing 4 *M* urea, 7.5% TEMED, 0.1% hydroxypropylmethylcellulose, and 3% 3–10 + 5–8 (50:50) ampholytes. Note the drop in baseline after all the sample components were detected. (Reproduced from Ref. 346 with permission.)

isoforms [289] by CIEF has been carried out in glass capillaries with their internal wall modified to minimize electroosmosis, in capillaries coated with linear polyacrylamide [322], and by CIEF with pressure mobilization [347]. The focusing patterns from normal subjects presented di, tri-, tetra-, penta-, and hexasialo transferrin glycoforms (Figure 39).

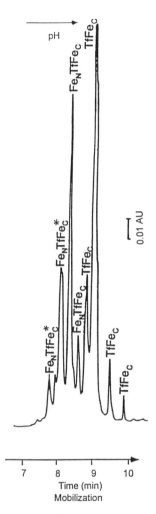

Figure 39 Analysis of human transferrin isoforms. On-line detection at 280 nm. Capillary IEF performed in glass capillaries, 100 μm × 14 cm coated with linear polyacrylamide; protein concentration was adjusted to 1 mg/ml with 2% Bio-Lytes 5–7. Focusing at 4 kV for 6 min. Anodic mobilization at 4kV. (Reproduced from Ref. 289 with permission.)

Capillary IEF was also used to observe changes in pI of transferrin molecules by the binding of iron in vitro using a concentration gradient imaging detection system [348]. First, transferrin was focused for 2 min in a 4 cm × 100 μm coated capillary by applying 3.5 kV. A 20 s pulse of iron solubilized in phosphoric acid was applied. With the aid of the concentration gradient detector (scanning the whole capillary) the formation and disappearance of focused transferrin bands was followed as the transferrin incorporated or released the iron ion.

Capillary IEF using pressure mobilization was used to estimate the pI values and characterize the metalloproteins conalbumin [347,349], transferrin, and metallothionein [347]. Using this method, it was possible to separate three major molecular forms of conalbumin (the egg form of transferrin) and transferrin, perhaps differing by the number of ferric ions they contained (zero, one, or two ferric ions). The iron-free forms of conalbumin and transferrin exhibited higher pI values than the diferric form of both proteins. This information could be used to assess the degree of iron saturation of serum transferrin. Metallothioneins are small acidic proteins that play an important role in cellular metal metabolism, since they have high affinity for heavy metals. Two isoforms from rabbit liver metallothionein were resolved.

Peptides

Mazzeo et al.[350] demonstrated the use of CIEF in the analysis of peptides. Using two well-characterized proteins (chicken and bovine cytochrome C), theoretical pIs based on amino acid sequence were first calculated for the expected tryptic peptides, and then compared with results obtained by CIEF in the presence of EOF. The pIs were determined by generating a standard curve with proteins of known pI values. With this method the correlation was only approximated, and while some peptides correlated near perfectly, others showed marked differences from theoretical values. Just as previously reported for proteins analyzed by CIEF in uncoated capillaries, the resolution and peak shape of peptides with acidic pIs were poorer than those for neutral and basic components, due to decreasing EOF as the analysis progressed. To eliminate this problem, a C8-coated capillary was used. Peak efficiency was increased in the C8 capillary and longer analysis times were also observed (C8 coated capillaries exhibited only about 40% EOF as compared to uncoated columns). Correlation between theoretical and estimated pI values was also poor in capillaries coated with C8. It is difficult to determine if the lack of correlation was due to the technique itself, or an example of the difficulties of predicting the pI of complex polypeptides.

Dyes

Departing from the traditional application of CIEF to the analysis of polypeptides, Caslavska et al. [351] studied the behavior of dyes under CIEF conditions. The aminomethylphenol dyes were found to be suitable as pI markers, exhibiting high solubility in ampholyte mixtures at their pI, long-term stability, no interaction with sample components or ampholytes, and high mobility around their pI. They did not significantly influence the EOF, which was used to transport the protein zones to the detector. Since these dyes are difficult to analyze by slab gels, this application is a clear example of the expanded role of electrophoresis when performed in a capillary format.

Protein Concentration and Dynamics of Interaction

An application of CIEF which exploits the concentration effect of the technique with the advantages of affinity interaction and the detection power of laser-induced fluorescence has been developed [309]. It is well known that a very important feature of many biological systems is specific recognition at the molecular level. Antibodies as a group are widely used for molecular recognition, e.g., affinity assays. This feature can be used by labeling an antihuman growth hormone antibody fraction with a fluorescent tag (tetramethylrhodamine-iodoacetamide) to detect the presence of growth hormone to a level of 0.1 ng per ml. Capillary IEF was performed in a 75 μm × 15 cm capillary coated with polyacrylamide and the analysis time was approximately 20 min. Due to the more acidic pI of the hormone, the unattached antibody was easily resolved from the antibody-growth hormone complex. Moreover, the system was able to discriminate protein variants, such as single- and double-deamidated growth hormone. By measuring the peak area of the complex after different incubation times the association time could be estimated. Through the use of CIEF it was determined that after only 2 min of incubation a saturation plateau was already reached.

FUTURE DEVELOPMENTS

The acceptance of CIEF as a routine analysis technique requires further developments in two major areas: (1) improved reproducibility, and (2) elimination or reduction of precipitation. We have discussed several strategies to improve the reproducibility of CIEF assays (e.g., buffered reagents, internal standards, etc.), but there is obviously room for improvement in this very important area. Reproducibility enhancement will increase the number of routine analyses performed by this technique; the elimination of precipitation (through the use of additives, optimization of analysis conditions, denaturing conditions, etc.) will expand the spectrum of application to include a greater variety of proteins. Most of the factors affecting resolution and reproducibility have been described throughout this chapter. Electroosmotic flow control is a key parameter to consider, both for two-step CIEF and for single-step CIEF. The former requires the elimination of EOF, and coatings with enhanced stability are highly desirable. For single-step CIEF, maintaining the EOF constant throughout the analysis is of prime importance.

Due to the disturbing effects of salt in the IEF process, it has become evident from our experience that in order to obtain good results, CIEF requires an easy, rapid (less than 5 min) method for on-line desalting. An alternative to desalting is to develop separation protocols that are more tolerant to salt content.

To date, all ampholytes used were developed for slab gel IEF, and there are several characteristics of these compounds that can be improved. First, ampholytes with low UV absorption would allow detection of proteins at low UV wavelengths (200–220 nm) and, therefore, much lower starting concentrations could be analyzed. The focusing effect of CIEF increases concentration by approximately

300 to 400-fold, and if detection were performed at 200 nm, a net increase in sensitivity of the same magnitude would be obtained (poor senstivity at 280 nm nearly offsets the gain in concentration). Detection at 200 nm would also allow the analysis of small peptides (at 280 nm peptides can be analyzed only if they possess aromatic amino acids). If the sample concentration could be reduced by 300 to 400-fold precipitation would also be less likely to occur. The combined effect of sample concentration and high power resolution would make CIEF the method of choice for the analysis of polypeptides.

Tables 3 through 7 provide guidelines for sample preparation, focusing and mobilization, and use of internal standards. These tables also provide typical analytical conditions for easy reference in developing a CIEF method.

Table 3 Guidelines for Sample Preparation

Ampholyte selection

- If the pI of the sample is unknown, or if contaminants with a wide pI distribution may occur, the first approach is to use broad-range ampholytes (e.g., Bio-Lytes 3–10) diluted to a solids concentration of 1–2%, preferably with a spacer (e.g., TEMED) to block the blind segment of the capillary.
- The concentration of the spacer should be optimized to cover only the capillary segment necessary to detect all sample components. For example, if the portion of the capillary beyond the detection point is 20% of the total capillary length, the spacer should be added to a concentration of 20% total solids. A typical ampholyte solution for this setup is then composed of 2% ampholytes + 0.4% spacer.
- Eliminate TEMED if the spacer is not required.
- To increase resolution, a narrow-range ampholyte mixture that spans the pH range of interest can be added to the broad-range ampholytes at different ratios, starting at 10% w/w and progressing until the desired resolution is achieved.
- Mixing ampholytes from different sources can also improve resolution.

Salt concentration

- Sample salt concentration should be as low as possible.
- When the salt content is unknown, a rule of thumb is that if the initial focusing current (see analysis parameters tables) is higher than 50 μA, the sample should be desalted. Obviously, this rule strongly depends on the properties of the protein(s) of interest. See the section on desalting protocols.

Table 4 Guidelines for Focusing

Typical focusing setup and conditions

Capillary	24 cm × 50 μm, coated (capillary's length and internal diameter depend on the type of mobilization to be employed; this capillary is optimal for ion addition mobilization)
Voltage	500–700 V/cm for 4 min
Anolyte	20 mM H_3PO_4
Catholyte	40 mM NaOH
Detection	280 nm
Capillary temperature	25°C
Sample temperature	25°C

Focusing events

- The charged ampholytes and proteins migrate under the influence of the electric field. If the initial focusing current is higher than 50 μA, the sample should be desalted.
- A pH gradient begins to develop, with low pH toward the anode (+) and high pH toward the cathode (−); the range of the pH gradient in the capillary is defined by the composition of the ampholyte mixture.
- At the same time, protein components in the sample migrate until a steady state is reached, at which point each protein becomes focused in a narrow zone at its isoelectric point.
- Focusing is accompanied by an exponential drop in current. Focusing is usually considered to be complete when the current has dropped to a level approximately 10% of its initial value and the rate of change approaches zero.
- The focusing process can be monitored by the movement of nascent protein zones past the detection point.
- If the whole capillary has been filled with sample and ampholytes, a drop in the baseline can be noticed after the focusing zones pass the detector point.
- The lack of focusing peaks may be due to low sample concentration or to high levels of EOF (the lack of mobilization peaks is a related phenomena).

Table 5   Guidelines for Chemical Mobilization

The following protocol is a guideline for performing CIEF with ion-addition mobilization in a 24 cm × 50 μm, linear polyacrylamide-coated capillary.

Focusing
Injection (at 100 psi)	15–40 s
Voltage	500–700 V/cm for 4 min
Anolyte	20 mM H_3PO_4
Catholyte	40 mM NaOH
Detection	280 nm
Capillary temperature	25°C
Sample temperature	25°C

Mobilization
Voltage	500–700 V/cm
Anolyte	20 mM H_3PO_4
Mobilizer	Zwitterionic mobilizer

Focusing time should be optimized to avoid overfocusing (which may lead to protein precipitation, and unnecessarily prolongs the analysis time), but also to prevent incomplete focusing (which reduces resolution, and may produce multiple peaks for the same sample component).

Chemical mobilization is affected by the type and concentration of competing ion used to mobilize, the presence of EOF, temperature, sample and ampholyte concentration, and voltage applied.

Table 6 Guidelines for Gravity Mobilization

The following protocol is a guideline to perform CIEF with gravity mobilization in a 30 cm × 75 μm, linear polyacrylamide-coated capillary.

Focusing
    Injection (at 100 psi)   10–30 s
    Voltage   500–700 V/cm for 4 min
    Anolyte   20 mM H_3PO_4 (full reservoir)
    Catholyte   40 mM NaOH (full reservoir)
    Detection   280 nm
    Capillary temperature   25°C
    Sample temperature   25°C

Mobilization
    Voltage   500–700 V/cm
    Anolyte   20 mM H_3PO_4 (full reservoir, e.g. 1.5 ml)
    Mobilizer   40 mM NaOH (partially filled reservoir,
            e.g. 0.5 ml)

Flow velocity during mobilization can be manipulated by changing the capillary length, and inner diameter, by adding hydrophilic polymers to the solution (e.g., methylcellulose, 0.05–0.2%), by adjusting the temperature of the reagents and capillary, and by controlling the volume difference between the injection and destination reservoirs.

Guidelines for single-step CIEF in the presence of siphoning (gravity)

In single step CIEF, the capillary is first filled with ampholytes. The sample is then injected, adjusting the loading parameters to obtain a sample plug of desired length.

Table 7 Guidelines for Using Internal Standards

Typical analysis conditions for markers introduced as separate injections:

Focusing

Capillary	24 cm × 50 μm, coated
Sample injection (at 100 psi)	15–40 s
Marker injection	2–5 psi*s each
Voltage	500–700 V/cm for 4 min
Anolyte	20 mM H_3PO_4
Catholyte	40 mM NaOH
Detection	280 nm
Capillary temperature	25°C
Sample temperature	2°C
Mobilization	Ion addition or hydraulic

Due to the small volume introduced when using separate injections (8–22 nl in this case, or 2–5% of the column volume) markers should be as concentrated as possible (e.g., diluted 1:1 v/v with the ampholyte mixture).

The internal standards can be added directly to the sample + ampholyte solution. In this case, the protocol does not change from those described previously. The pI markers should be selected to bracket the sample components of interest, without interfering with the sample pattern, and their concentration adjusted to produce detector signals in the same range as the sample components.

It is more practical to introduce the pI markers as a separate injection. The main benefit of this approach is the flexibility to use different markers with the same sample. The main drawback is that since each marker displaces a small amount of sample from the interior of the capillary, sensitivity may be compromised if too many markers are used.

Markers should be prepared in the same ampholyte mixture used for the sample.

Caution: The linearity of the pI marker calibration depends on the linearity of the ampholytes used. If narrow-range ampholytes are used, and any of the markers pI is outside this added range, the pI determination of the sample components will not be accurate.

7

Sieving Separations

The performance of electrophoresis in narrow bore capillaries obviated most of the functions of gels in electrophoresis, e.g., elimination of convection (through rapid dissipation of Joule heat) and reduced diffusion (through short analysis time). However, another important feature of gels is their capability to actively participate in the separation process by providing a sieving medium that differentially affects the migration velocity of sample components according to molecular size. Macromolecules such as nucleic acids and SDS-protein complexes exhibit no significant mobility differences during free zone electrophoresis, and require the presence of an interactive, sieving separation matrix.

Size-based analysis of SDS-protein complexes in polyacrylamide gels (SDS-PAGE) is the most common type of slab gel electrophoresis for the characterization of polypeptides, and SDS-PAGE is one of the most commonly used methods for determination of molecular masses of proteins [352]. The uses of size-based techniques include purity determination, molecular size estimation, and identification of posttranslational modifications [353,354]. Some native protein studies also benefit from size-based separation.

Due to the importance and broad spectrum of applications of sieving separations, adapting gels to the capillary format has been attempted by many groups [355–357]. Unfortunately, there are several technical difficulties that have limited the use of gel-filled capillaries, such as bubble formation, contamination after repeated runs and, for polypeptides, poor detection sensitivity. A rapidly expanding alternative to gel-filled capillaries is the use of polymer solutions [358]. Advances in both of these methods are described below.

In this chapter, we refer to gel-filled capillaries as those containing cross-linked matrices that are fixed to the interior of the capillary, that is, they are not replaceable. Replaceable matrices are referred to as polymer solutions. Gels are typically polymerized in situ, whereas polymer solutions are pumped into the capillary and are usually replaced between each injection. This technique has been variously termed entangled polymer CE, nongel sieving, and dynamic sieving, and the polymer solutions have been referred to as replaceable gels and physical gels. A list of sieving media used for size-based separations in capillary electrophoresis is presented in Table 1.

ANALYSIS OF NATIVE PROTEINS

Native proteins consisting of varying numbers of identical subunits or protein conjugates made up of monomers joined by cross-linking agents may be difficult to resolve by free zone electrophoresis of cIEF, but they can be easily separated by sizing methods. In some instances, it is desirable to maintain the tertiary and quaternary structures of proteins and protein aggregates, which are lost when the polypeptides are denatured. In these cases, sieving of native proteins can be performed using either gel-filled capillaries [357,359] or poly-

Table 1 Sieving Media Used for Size-Based Protein Separations

Sieving media	References
Gels:	
Cross-linked acrylamide	225, 235, 237
Linear acrylamide	212, 213, 221
Polymer solutions	
Linear polyacrylamide (LPA)	215, 218, 222, 229
Polyethylene glycol,	215, 219, 221, 228, 232, 233, 234
Polyethylene oxide (PEG, PEO)	
Pullulan	224
Poly(vinyl alcohol) (PVA)	227
Dextran	217, 221, 226, 232
Low melting point agarose	230

mer solutions [360]. Size-based analysis of native bovine serum albumin (BSA), is shown in Figure 1.

Low-melting agarose has been proposed as a sieving medium for capillary electrophoresis [359]. This material can be introduced into the capillary by pressure at a temperature above its melting point of 25.6°C (e.g., 40°C), then induced to form a gel by dropping the capillary temperature below the melting point (e.g., to 20°C). Following the separation, the capillary temperature can be raised and the gel extruded from the capillary by pressure, and then replaced with fresh uncontaminated agarose prior to the next injection. Unfortunately, the pore size of agarose gels is too large to provide sufficient sieving strength for most proteins.

ANALYSIS OF SDS-PROTEIN COMPLEXES

Sodium dodecylsulfate binds to water-soluble proteins in an approximately stoichiometric fashion to the polypeptide chain, with one SDS molecule bound per two amino acid residues [361]. Therefore SDS-protein complexes will possess the same charge-to-mass ratio independent of polypeptide chain length, assuming that the contribution of the charged amino acid side chains is low relative to that of the surfactant phosphate groups. As expected, SDS complexes of proteins with molecular masses greater than 10,000 kDa exhibit identical mobilities in free solution [362], although proteins which are not fully complexed with SDS may exhibit variable mobilities and may be resolved into multiple species [363]. An important advantage of using SDS to denature polypeptides is the solvation power of the detergent. This property allows the study of proteins that easily precipitate under most other conditions (e.g., membrane proteins). Use of SDS permits the study of complex mixtures without the need for extensive sample clean-up.

Size-based analysis by capillary electrophoresis provides similar information and comparable limits of detection to those obtained by SDS-PAGE with Coomassie blue staining [355,364]. Although CE analyzes one sample at a time, total analysis time for multiple samples is shorter for CE than for a 16-lane slab gel [365]. The

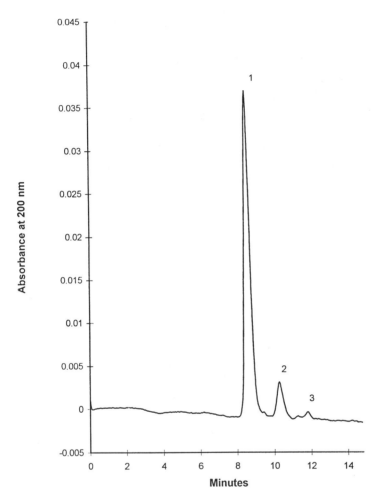

Figure 1 Separation of aggregates of native bovine serum albumin by entangled polymer sieving capillary electrophoresis. Peak 1, monomer; peak 2, dimer; peak 3, trimer. Analysis conditions: capillary, 24 cm × 50 μm, polyAAEE-coated; injection, 5 psi*s; buffer, 50 mM sodium phosphate (pH 2.5) + 5% 35 kDa PEG; 10 kV with positive to negative polarity; capillary temperature 20°C.

performance of both electrophoretic techniques for the analysis of polypeptides is superior to size exclusion chromatography.

Protein separations by capillary electrophoresis are often negatively affected by sample-capillary wall interactions, and often require additives or surface modifications to eliminate undesirable interactions (see Chapter 4). This problem is minimal or nonexistent for SDS–protein complexes, since untreated silica possesses negative charges at pH above 2–3, and the SDS–protein complex is also anionic at almost any pH, resulting in electrostatic repulsion, and thus eliminating protein adsorption. Nevertheless, in cases where EOF limits resolution or introduces migration time variations, internally coated capillaries can be used [384]. Capillaries coated with linear polyacrylamide through C–Si bonds were found to be more stable than capillaries coated through siloxane groups [366–368]. Uncoated capillaries have been used with a linear polyacrylamide sieving matrix providing sufficient viscosity (>100 cP) to prevent extrusion of the sieving medium from the column by EOF [352]. In some instances, the sieving matrix acts as a surface coating [369].

SDS-protein complexes have been resolved using gel-filled capillaries or polymer solutions. Next, we describe some of the advantages and disadvantages of each of these methods.

Gel-Filled Capillaries

As stated above, we consider as gels those matrices that are fixed to the interior of the capillary, and thus are not replaced between analyses. The first described use of gel-filled capillaries for analysis of SDS-denatured proteins was in 1983 [357]. Since then, most reports employed either of two types of gels: polyacrylamide cross-linked with bis-acrylamide [356], and linear polyacrylamide. Both gels are polymerized in situ, since their high viscosity precludes pumping them into the narrow bore column. Since one of the drawbacks of gel-filled capillaries is poor lifetime, several studies have aimed to increase the useful time of the columns. It was found that the lower the degree of cross-linking, the longer lifetime of the column [370]. Thus, linear polyacrylamide gels (zero cross-linking) were introduced. These gels were also more compatible with high electric fields than cross-linked gels [355]. Unfortunately, even gels with zero cross-linking were able to be used for only 20 to 40 runs.

The efforts to adapt gels to the capillary format were due their high resolving power compared to polymer solutions [355]. Disadvantages of gel-filled columns include a short lifetime, low reproducibility, and poor detection sensitivity due to high UV absorption of the gel matrix. Protein detection in gels is usually accomplished at 280 nm, but the extinction coefficients of proteins are 20–50 times larger at 214 nm. At 214 nm there is also less variability in the intensity of absorbance of proteins [370]. Another drawback of gel-filled capillaries is that the composition of the gel-filled capillary cannot be changed, which is a disadvantage in procedures such as the generation of Ferguson plots (see below).

Polymer Solutions

The advantages of using solutions containing polymers as additives to achieve size-based separations include increased reproducibility (since the capillary's content is replaced at each run), increased capillary lifetime, the possibility of using polymers with low absorption in the 200–220 nm range, and ease of storage and handling. Separation parameters are simple to optimize by changing polymer type and concentration, buffer pH, viscosity, and conductivity. As stated above, the main drawback of polymer solutions is that resolution is not as high as that obtained with gel-filled capillaries.

Several types of polymers have been shown to be suitable for separation of a broad-molecular-weight range of polypeptides. Care should be exercised when selecting polymers and optimizing analysis conditions [371], since separation parameters such as temperature do not affect all polymers equally [372–374]. Resolution also depends on the type, size, and concentration of polymer used. Under optimized conditions, polypeptides differing by as little as 4% in molecular mass can be resolved [369]. Early reports on the use of polymer solutions for the analysis of SDS complexes included dextran and polyethylene glycol (PEG) [370]. Both of these polymers are practically transparent at 214 nm, and thus greatly improved detection over polyacrylamide gels. Migration time (MT) reproducibility is of prime importance in this technique, since migration times are used to estimate the molecular size of proteins. The use of a replaceable matrix increases MT reproducibility, and RSD values as low as 0.3% were obtained using dextrans (similar values were obtained for PEG matrices). Using polymer solutions, the life of the column was also extended to over 300 analyses.

A list of some polymers used as sieving matrices is displayed in Table 1. One important consideration when selecting a polymer is the viscosity of the final solution. In most cases, low viscosity is desired for easy replacement of the capillary content in-between analyses, but uncoated capillaries will exhibit higher EOF with lower viscosity polymers [373]. Resolution achieved using polymer sieving capillary electrophoresis is comparable with that obtained using a 12%T polyacrylamide slab gel.

Generally, one of three mechanisms are considered to explain size-base separations [354]:

1. The Ogsten model implies true sieving, and when this mechanism is in effect the pore size of the matrix should be in the same range as the hydrodynamic radius of the sample molecules.
2. The reptation model is used to explain the migration of long, polyionic molecules (e.g., nucleic acids) with a larger molecular radius than the size of the matrix pores.
3. The third model is used when high electric fields are involved, since reptation with stretching may occur.

Guttman [354] suggested that the mechanism of separation for SDS-protein complexes analyzed in solutions containing polyethylene oxide (PEO) is reptation with stretching. Plots generated under various electrophoresis conditions and different polymer characteristics and concentrations were constructed and analyzed, resulting in the elimination of the Ogsten and reptation mechanisms. According to the Ogsten model, a linear relationship is expected for molecular mass and the logarithm of mobility. As shown in Figure 2, this was not true for any of three different molecular weights of PEO used. The increased curvature of the plots obtained with the high-molecular-weight PEO suggested a possible reptation mechanism. To

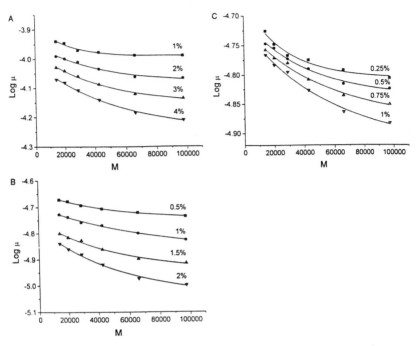

Figure 2 Relationship between the logarithm of electrophoretic mobility of different sized SDS-protein complexes and the protein molecular weights. Sieving matrices used: (A) 100 kDa polyethylene oxide (PEO), (B) 300 kDa PEO, and (C) 900 kDa PEO. Numbers in the plots correspond to the actual concentration of the sieving polymer solution. Test mixture contained α-lactalbumin (14.2 kDa), soybean trypsin inhibitor (21.5 kDa), carbonic anhydrase (29 kDa), ovalbumin (45 kDa), bovine serum albumin (66 kDa), and phosphorylase B (97.4 kDa). (Reproduced from Ref. 354 with permission.)

differentiate these two regimes, the logarithm of mobility of the solute was plotted as a function of the logarithm of the solute molecular mass. A slope value of −1 should be obtained when pure reptation is the mechanism of separation. As indicated in Figure 3, the values of slopes obtained were much lower (−0.07 to −0.18) than those expected for pure reptation. Three arguments favored a separation mechanism based on reptation with stretching (Figure 4): (A) lower-

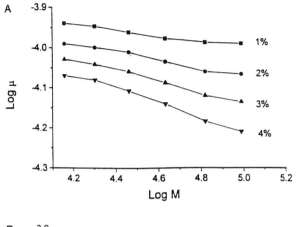

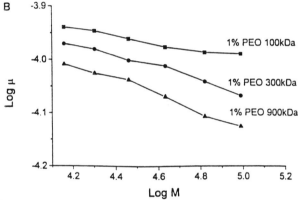

Figure 3 Double logarithm plots of the electrophoretic mobility and protein molecular mass (*M*) for a protein test mixture. (A) Sieving matrix: 100 kDa PEO, numbers in the plots correspond to the actual sieving matrix concentration. (B) Sieving matrix : 1% PEO (100 kDa), 1% PEO (300 kDa), and 1% (900 kDa). (Reproduced from Ref. 354 with permission.)

than-expected slope values for log mobility vs. log molecular mass plots, (B) nonlinear increase of mobility with increasing electric field strength, and (C) variation in extrapolated mobility at zero polymer concentration when the mass of the polymer was increased. According to this model and to the plots obtained, the sieving strength of a

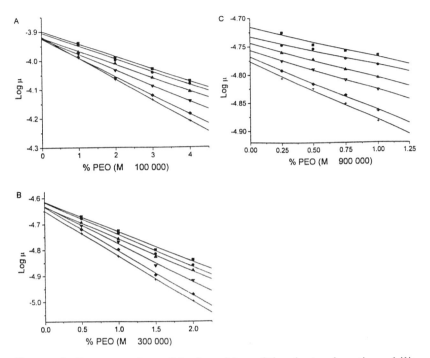

Figure 4 Ferguson plots of the logarithm of the electrophoretic mobility and the sieving polymer concentrations of different sized polymer. (A) 100 kDa PEO, (B) 300 kDa PEO, and (C) 900 kDa PEO. The sample used is the same protein mixture described in Figure 3. (Reproduced from Ref. 354 with permission.)

polymer solution can be increased by increasing the concentration of a given size polymer, and/or by increasing the length of the polymer used (maintaining the concentration constant).

Karim et al. [375] studied the size and molecular weight distribution of dextrans and their effect on resolution of SDS–protein complexes. Dextrans of higher molecular-weight and narrow size distribution were found to produce higher resolution, specially for the larger proteins studied (BSA and phosphorylase B, 66250 and 97400 Da, respectively). Unexpectedly, the solutions containing low-molecular-weight dextrans (1270 Da, and 5220 Da) were also able to resolve these two proteins. This is of interest in regard to the separation mechanism, since such oligosaccharides probably do not form a network structure. In a related study, Simo-Alfonso et al. [373] compared the behavior of sieving solutions containing poly(vinyl alcohol) (PVA), and found that optimal concentration of the polymer showed a strong dependence on the i.d. of the capillary used. The authors characterized several polymer solutions by using global resolution (R_{sg}, "the product of each individual resolution divided by the average of individual resolutions"). This permitted an easier comparison, especially when polymers exhibited marked differences in resolving power at different sample size intervals. For a 50 μm capillary, R_{sg} was about five times higher than for a 75 μm column, and the optimal PVA concentration shifted from about 4 to 6%. The same effect was observed if the number of plates was used instead of R_{sg}. For a 25 μm capillary, the highest R_{sg} obtained was at a concentration of about 1%, which is below the entanglement threshold of PVA (3%). This finding is of significance because it is generally believed that polymer solutions exhibit sieving power only if their concentration is near or above the entanglement threshold. According to DeGennes [376], polymer solutions exhibit two types of regimes: one in which their concentration is low and the individual polymeric chains are isolated from each other, and a second regime when the polymer concentration is increased and the polymeric chains start to overlap. As the polymers begin to overlap (entanglement threshold) they become densely packed. Polymers in this regime form a dynamic porous structure that can actively participate in the separation process, in a manner analogous to cross-linked gels. Barron et al. [377,378] demonstrated that sieving of DNA fragments occurs in extremely dilute hydroxyethylcellulose (HEC) solutions (≤0.002%), which is below its entanglement threshold of 0.35%. Their findings can be fully explained through the existence of a transient entanglement coupling mechanism. Large DNA filaments will screen a large number of HEC coils (which are in average 10 times smaller than the DNA fragments used) and become entangled, whereas short DNA filaments could not form this coupled system. Still, this theory does not explain Karim and Righetti's findings, since SDS–protein complexes are much shorter than long filaments of DNA. Righetti proposed an additive mechanism of separation: (1) SDS–protein complexes and comicelles are in reality much longer than a polypeptide alone, and (2) PVA chains might have a unique entanglement regime due to extensive hydrogen bonding.

Ferguson Analysis

Molecular-weight estimation of SDS–protein complexes is usually performed by comparing the migration times of sample components with those of proteins of known molecular weights (MW). The curve obtained with protein standards may introduce errors in the estimation of MW if the binding of detergent by the protein is anomalous (e.g., membrane proteins, glycoproteins, highly basic proteins). Since detergent binding directly affects protein mobility by changing the mass-to-charge ratio, the MW discrepancy originates from differences in free solution mobilities of the different polypeptides [364]. One way to avoid such errors is to construct Ferguson plots. These are made by measuring the migration times at different polymer network concentrations, and constructing a universal calibration curve [379] by plotting the logarithms of the relative migrations as a function of polymer concentration. According to Ferguson, the logarithm of the protein's mobility varies linearly as a function of the gel concentration employed. The slope of this mobility line yields a parameter called the retardation coefficient (K_r), which is proportional to the square of the radius of the protein. Universal standard curves are constructed by plotting the logarithm of known protein MW as a function of the square roots of the retardation coefficients. The slope of the curve represents the retardation coefficient while the intercept at zero polymer concentration corresponds to the free solution mobility of a protein. An intercept located at a point other than zero on the concentration scale at zero polymer concentration is an indication of differences in free solution mobilities of the SDS-protein complexes. Proteins with similar molecular radii show the same slope, independently of where they intersect on the concentration axis. Ferguson analysis for traditional SDS-PAGE is extremely time consuming, especially because the analysis should be performed using at least six different gel concentrations. Consequently, this method of analysis was practically abandoned until the use of CE with replaceable polymer networks made the Ferguson analysis more feasible. A Ferguson plot can be generated automatically by CE using different dilutions of the sieving buffer [379,380]. Ferguson plots are also used to elucidate separation mechanisms [354].

PRACTICAL CONSIDERATIONS IN ANALYSIS OF SDS-PROTEIN COMPLEXES USING POLYMER SOLUTIONS

Commercial kits for analysis of SDS-protein complexes using entangled polymer sieving systems are currently available from Bio-Rad Labortories, Beckman Instruments, and the Applied Biosystems Division of Perkin Elmer. This discussion is based on the authors' experience with the replaceable polymer sieving system developed in their labs at Bio-Rad. This employs a proprietary hydrophilic sieving polymer in 0.4 M Tris borate buffer (pH 8.5) containing 0.1% SDS. The chain length and concentration of the sieving polymer was formulated to provide resolution of SDS-protein complexes over a molecular-weight range of 14–200 kDa. The buffer also contains a low concentration of the polymer modified by introduction of charged functions. The combination of the high-viscosity sieving polymer and the cationic modified-polymer additive serves to reduce the electroendosmotic flow in an uncoated capillary to less than 5 × 10^{-5} cm^2 V^{-1} sec^{-1}. Because proteins which have been complexed with SDS are strongly anionic, they do not adsorb to the capillary wall under the alkaline run conditions, allowing analyses to be performed in uncoated capillaries in the absence of significant EOF.

Sample Preparation

Preparation of samples for SDS-CE analysis is essentially the same as for SDS-PAGE. Protein samples are diluted 1:1 in a Tris-HCl + SDS (pH 9.2) sample preparation buffer; the final buffer concentration after dilution is 50 mM Tris HCl + 0.5% SDS. If the proteins are to be analyzed under reduced conditions, an appropriate reducing agent should be added, e.g., β-mercaptoethanol (final concentration 2.5%) or dithiothreitol (15 mM). If the analysis is to be used for estimation of protein molecular weight, an internal standard can be added to the sample for use in the normalization of protein migration times. The Bio-Rad CE-SDS Kit employs benzoic acid as the internal standard, added to a final concentration of 50 mg/ml. After mixing the sample, buffer and internal standard, the mixture should be heated at 95–100°C for 10–12 min to complex proteins with the SDS. It is our experience that heating in a water bath is necessary; use of a contact heating block is not always effective and may result in reduced separation efficiency.

The presence of salt in the sample will interfere with the injection process, and highest sensitivity will be obtained if the sample salt concentration is less than 50 mM. Samples containing higher salt or buffer concentrations should be desalted by dialysis or ultrafiltration. If there is any particulate material remaining in the sample after heating, it should be removed by centrifugation or filtration.

Analysis Buffer Preparation

Entangled polymer sieving buffers are quite viscous, and bubbles are frequently trapped in the bottom of the buffer reservoir vials when the analysis buffer is pipetted into them. In this situation, the capillary orifice and high voltage electrode will not contact the buffer, resulting in erratic current and failed analyses. To prevent this, the buffer vials should be centrifuged for at least 2 min in a microcentrifuge immediately prior to installing them in the CE instrument.

Capillary Preparation

The Bio-Rad CE-SDS analysis buffer is designed for use with uncoated capillaries, and no prior capillary conditioning is required. However, the capillary should be purged with acid and base wash solutions before replenishing the run buffer before each injection. When using the recommended 24 cm × 50 μm capillary dimensions, a 90 s purge with 0.1 N NaOH followed by a 60 s purge with 0.1 N HCl should be used. These purge cycles serve to sweep residual buffer and any remaining sample components from the capillary. Fresh run buffer is then introduced with a 120 s purge step. The viscosity of the CE–SDS analysis buffer is approximately 43 centipoise, and the purge times given are adequate for purge pressures of 100 psi. If the CE instrument employs lower purge pressures or if different capillary dimensions are used, purge times should be modified appropriately.

Because of the high viscosity of the entangled polymer solutions, buffer can be retained on the outer surfaces of the capillary and electrodes after the replenishment step, resulting in carryover of buffer into the sample solution during injection. This will reduce injection efficiency and compromise sensitivity. To prevent this, the capillary and electrode surfaces should be washed between replenishment and injection. A convenient method of doing this is to add one or two purge steps prior to injection in which the capillary is immersed in a wash solution (water or diluted sample preparation buffer) without application of pressure. Separate vials of wash solution should be used for each rinse or "dip" cycle.

Injection

In CZE, electrophoretic or electrokinetic injection is usually not the preferred injection mode because of the phenomenon of electrophoretic bias: sample ions of low mobility will migrate more slowly in the injection process and therefore be at lower relative concentrations in the starting zone. In the case of SDS–protein complexes, all sample components will have approximately the same mass-to-charge ratio because of the constant charge density of SDS on the protein. Therefore, all SDS–protein complexes will be loaded with the same efficiency using electrokinetic injection. The high ionic strength of the analysis buffer (0.4 M Tris-borate) provides a stacking effect, decreasing starting zone width and increasing zone concentration. So in most cases, electrokinetic injection is the preferred mode for this technique. However, if the sample contains appreciable salt concentration and it is not practical to desalt the sample, pressure injection may be used. The high viscosity of the run buffer requires sufficient injection times to introduce enough material. For example, in the case of a 24 cm × 50 μm capillary, an injection of 12 s at 5 psi is necessary to inject a 0.3 mm sample zone. If sample salt concentration is greater than 50 mM, even pressure injection will not provide satisfactory sensitivity.

Detection

The great advantage of entangled polymer systems is their transparency in the low UV range. However, the absorbance of the buffer components (e.g., Tris) and the SDS contribute appreciable background signal below 210 nm. Detection at 220 nm reduces background interference without significant loss in protein response.

Analysis Conditions

Operation at a field strength of 625 V/cm provides satisfactory resolution with short run times (Figures 5 and 6); typical current is approximately 20 μA using a 24 cm × 50 μm capillary. The capillary should be thermostatted close to ambient temperature (e.g., 20°C) for good reproducibility.

It is extremely important not to expose the capillary tips to drying conditions when using entangled polymer sieving buffers. The sieving polymer will precipitate, plugging the capillary. If the capillary becomes plugged, it can sometimes be regenerated using the procedures described in Chapter 5, but these procedures are less successful with polymer-containing buffers. It is highly recommended that the capillary be immersed in buffer after use, or preferably purged with at least five capillary volumes of water.

Performance

To be considered as a useful alternative to conventional SDS-PAGE, an SDS-CE system must provide satisfactory resolution, precision, and throughput. In addition, it should provide reliable molecular weight estimates and quantitative information.

Resolution

A separation of protein standards in the molecular-weight range of 14–200 KDa using a 24 cm × 50 μm capillary operated at 15 kV is shown in Figure 5. Resolution is similar to that obtained with a 12% T, 2.6%C SDS-PAGE gel. Resolution can be improved somewhat by using a longer capillary at the expense of longer analysis times; in our experience, the entangled polymer sieving systems are less reliable using longer capillaries and runtimes. Resolution can be increased by operation at lower field strengths, also at the expense of analysis time.

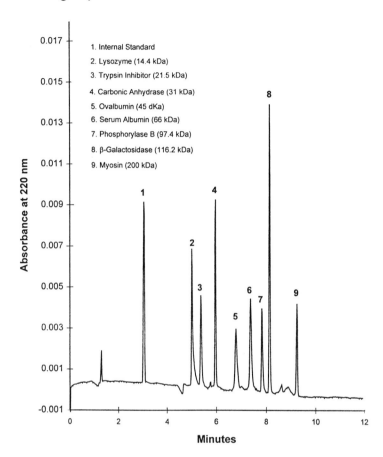

Figure 5 Separations of protein standards using the Bio-Rad CE-SDS Kit. Conditions: Capillary, 24 cm × 50 μm, uncoated; injection, 4 kV for 10 s; buffer, 0.4 *M* Tris borate (pH 8.3) + 0.1% SDS + proprietary sieving polymers; run voltage 15 kV with negative to positive polarity; capillary temperature 20°C.

Precision

Precision of migration times and peak areas for the eight protein standards using electrokinetic injection are presented in Table 2. Migration time precision was about 0.5% RSD and peak area precision is about 1–3%. Peak area precision using pressure injection was comparable (data not shown).

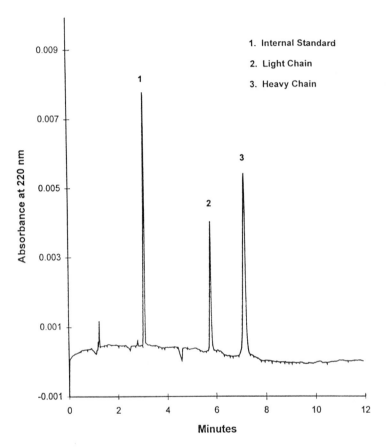

Figure 6 Separation of IgG monoclonal antibody reduced with dithiothreitol. Analysis conditions same as Figure 5.

Throughput

Injection-to-injection time, including capillary purge sequence, separation, and integration, is about 17 min. The time to prepare samples for CE-SDS is comparable to that for SDS-PAGE, so throughput is approximately 18 min/sample, of which only sample preparation is hands-on time. This is compared with throughput estimates for SDS-PAGE in Table 3.

Table 2 Analytical Precision of SDS-Protein Separations Using Polymer Sieving Capillary Electrophoresis[1]

Protein	Migration Time %RSD	Peak Area %RSD
Lysozyme	0.35	1.51
Trypsin inhibitor	0.40	0.85
Carbonic anhydrase	0.45	1.24
Ovalbumin	0.55	1.57
Serum Albumin	0.56	3.30
Phosphorylase B	0.55	3.32
b-Galactosidase	0.59	3.05
Myosin	0.66	1.76

[1]$n = 10$; electrokinetic injection for 10 s at 10 kV. All data obtained using the CE-SDS Protein Kit from Bio-Rad Laboratories on a BioFocus 3000TC automated capillary electrophoresis system.

Table 3 Performance Comparison of Polymer-Sieving CE and SDS-PAGE[1]

	Throughput (min/sample)	Concentration Sensitivity (µg/ml)	Mass Sensitivity (ng applied)
Polymer sieving CE	18	0.5	0.02
SDS-PAGE (Coomassie staining)	10	2–20	100–1000
SDS-PAGE (zinc staining)	6	0.2–2	10–100
SDS-PAGE (copper staining)	6	0.4–2	20–100
SDS-PAGE (silver staining)	10	0.04–0.2	2–10

[1]Assumes 50 µl volume applied to each lane of a 10-lane precast 12%T, 2.6%C SDS-PAGE gel; time includes electrophoresis, staining, and densitometric scanning. SDS-PAGE data obtained from the 1997 Bio-Rad Life Science Research Products Catalog, pages 106–110.

Quantitation

The ability to acquire quantitative information on protein concentration is considered a major advantage of CE compared to SDS-PAGE, in which staining response has only narrow linear ranges, depends on operator technique, and is subject to batch-to-batch variability of the stain. In contrast, protein response using polymer sieving CE with UV detection at 220 nm is linear over three orders of magnitude (Figure 7) and quite reproducible (Table 2). In this system, the detection limit for carbonic anhydrase (S/N = 3) is 0.5 μg/ml.

Estimation of Molecular Weight

Using the polymer sieving system described above, the log of protein molecular weight is correlated with migration time (Figure 8). Molecular weight can be determined directly by comparison to migration times of standard proteins. However, small variations in migration times can introduce significant error in the calculated molecular weight value. More reliable estimates may be obtained by normalizing the migration times of protein standards and samples to that of an internal standard such as benzoic acid. Correlation of

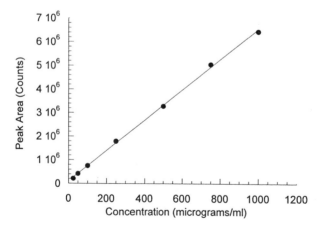

Figure 7 Peak area vs. concentration of carbonic anhydrase. Analyses performed using the Bio-Rad CE-SDS Kit with same conditions as Figure 5.

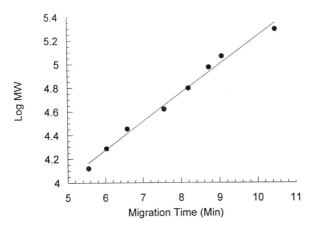

Figure 8 Log protein molecular weight vs. migration time. Data obtained from the proteins shown in Figure 5 using the same analysis conditions.

migration time with molecular weight for 34 proteins is shown in Figure 9, and the agreement between molecular weight values reported in the literature and those determined by SDS–CE is shown in Figure 10.

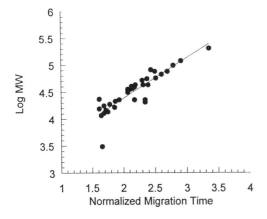

Figure 9 Log protein molecular weight vs. migration time. Migration times determined for 34 proteins from commercial sources with the Bio-Rad CE-SDS Kit using the conditions described in Figure 5. Molecular-weight values obtained from the literature.

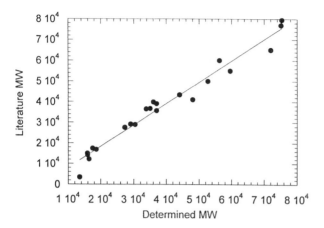

Figure 10 Molecular-weight estimates obtained for several proteins obtained using the Bio-Rad CE-SDS Kit compared with values for the same proteins reported in the literature.

APPLICATIONS

Collagens

Collagens are a family of structural proteins consisting primarily of two polypeptides (α_1 and α_2) of about 1000 amino acids in length and associated in a triple helix. Deyl and Miksik [244] employed entangled polymer CE with 4% linear polyacrylamide to separate collagen-SDS complexes. Collage Type I α_1 and α_2 chains were resolved, as well as β_{11}, β_{12}, and γ-chain polymers. Separation of glycated collagens produced by incubation with glucose was also achieved.

Cereal Proteins

Werner et al. [381] used a linear polyacrylamide entangled polymer sieving system to separate wheat endosperm proteins extracted from crushed wheat seeds with 1% SDS + 1% 2-mercaptoethanol. This procedure extracted both gliadins and reduced glutenin subunits. Capillary electrophoresis using the SDS-entangled polymer medium supplemented with 5% methanol and 3.8% glycerol yielded separation profiles similar to those obtained with conventional SDS-PAGE, and permitted differentiation of several wheat cultivars. It has long been known that SDS-PAGE yields molecular-weight estimates for wheat glutenin subunits which are lower than those obtained by ultracentrifugation or calculated from amino acid sequence data. Similar anomalous molecular-weight values were obtained by Werner [381] using the entangled polymer system described above (but without methanol and glycerol additives). Ferguson-type analyses were performed by determining the mobility of standard proteins and glutenin subunits in a series of sieving media containing variable concentrations of linear polyacrylamide. The slopes of mobility vs. polymer concentration were used to construct a plot of retardation coefficients vs. molecular weight of protein standards, and the molecular weights of the glutenin subunits were estimated by interpolation. Results agreed with M_r values obtained from sequence data, and suggested that the anomalous mobility of glutenin subunits was due to incomplete SDS binding to the subunits.

Milk Proteins

Quantitative analysis of proteins in bovine whey was carried out by Cifuentes et al. [382] using a polyacrylamide-coated capillary and an electrophoresis buffer containing 0.1% SDS with 10% polyethylene glycol 8000 as the sieving polymer. Concentrations of β-lactoglobulins A + B, α-lactalbumin, and bovine serum albumin agreed closely with values obtained by reversed phase HPLC, and analysis time was reduced fourfold compared to the gradient chromatographic method.

Kinghorn et al. [193] used a commercial linear polyacrylamide-based polymer sieving system to separate SDS complexes of proteins in liquid bovine whey and reconstituted whey protein concentrate. Pretreatment of the acid whey sample by dilution in 2% SDS or by ultrafiltration was necessary to prevent precipitation of the SDS as potassium dodecylsulfate. The levels of α-lactalbumin and β-lactoglobulin determined by this method agreed satisfactorily with other methods (native- and SDS-PAGE, size exclusion-, ion exchange-, and affinity-HPLC).

Recombinant Proteins

SDS–PAGE is widely used in the biopharmaceutical industry to monitor the purification and to estimate the purity of recombinant protein therapeutics. However, SDS-PAGE is a labor-intensive and and (at best) semiquantitative technique which is not ideally suited for high-throughput analysis in commmercial laboratories. Kundu et al. [383] have evaluated a commercially available entangled polymer sieving kit for capillary electrophoresis of SDS-proteins and compared the results with SDS-PAGE with Commassie Brilliant Blue (CBB) using a 72 kDa viral-CKS fusion protein. Protein purity was assessed from densitometric scanning of CBB-stained gels and from integration of electropherograms monitored at 220 nm. The authors demonstrated that purity levels determined by the two methods were in good agreement, and that SDS-PAGE quantitation was limited by the nonlinearity of CBB staining at low concentration. Molecular mass estimates obtained using MALDI-TOF mass spectrometery, SDS-CE, and SDS-PAGE were found to differ from the theoretical molecular mass by 100, 600, and 1400 mass units, respectively.

8
Concluding Remarks

The goal of this book has been to present the broad spectrum of separation methods for protein analysis by capillary electrophoresis and to provide some guidelines in their practical application. These methods include separations based on mass and charge, isoelectric point, molecular sieving, and affinity interactions. In fact, CE separation modes analogous to most existing electrophoretic and some chromatographic techniques are currently available. In many cases, resolving power exceeds that of conventional methods by virtue of the excellent heat dissipation of fused silica capillaries and the plug-flow characteristics of EOF. In spite of these advantages, CE is now only beginning to be established as a routine analytical tool in the protein chemistry lab. This can be attributed to limitations in reproducibility, sensitivity, and throughput. It seems appropriate at the conclusion of this work to address these issues with an eye to the future evolution of protein CE.

Reproducibility

Poor reproducibility of separation patterns and peak response has been a concern for capillary electrophoresis in general and protein CE in particular since its early development in the 1980s. It was widely appreciated that protein-wall interactions were largely responsible for variable EOF and the resulting variability in migration times and peak areas. As a consequence, the development of means to control or eliminate wall interactions continues to be a major research effort in CE. Pioneering efforts by Hjertén, Regnier, Schomburg, El Rassi,

Englehardt, and others lead to the development of separation chemistries and capillary coatings which provided successful protein separations in many applications. At this writing, over a dozen coated capillaries are commercially available. However, there is clearly room for improvement as commercial coated capillaries still do not provide satisfactory lifetime under harsh analysis conditions, e.g., high pH. It is hoped that improved bonding conditions and coating structures as well as improvement in the quality of the fused silica will provide more stable capillaries with reproducible performance. An alternative strategy is the use of rapid in situ coating methods with which a user can replace lost coating periodically, recovering lost performance and greatly extending capillary lifetime at minimal cost.

CE performance is also affected by the sample matrix. The excitement generated among researchers by million-plate separations with pure standards often evaporated when comparable performance could not be achieved with real samples under routine analytical conditions. Matrix affects are common in chromatography, but they are more serious in CE where sample ionic strength and viscosity can profoundly affect injection efficiency and quantitative precision. Acceptance of CE as a useful technique will therefore depend not only on improvements in instrumentation and separation chemistries, but also on improved understanding of sample matrix effects, simplified rapid sample handling techniques, and development of suitable internal standards.

Sensitivity

Sensitivity in CE is typically limited by the short path lengths imposed by on-tube detection with narrow-bore capillaries. On-tube absorbance detection, even at short wavelengths, may not provide the sensivity for detecting minor sample components such as impurities and degradation products which may be readily detected on silver-stained gels. Extended-pathlength cells (e.g., bubble cells, Z-cells) have been introduced to enhance sensitivity, but these can compromise resolution. Fluorescence detection with laser-based sources has dramatically increased sensivity for nucleic acid and carbohydrate analysis using and intercalating dyes and labeling techniques. Unfortunately, derivatization of proteins to introduce fluorophores typically employs amine-specific reagents, and incomplete reaction generates multiple products. Their variation in mass/charge ratios produces peak broadening and efficiency loss. Fluorescent labeling techniques which preserve the high efficiency of CE separations represents a fertile area for continued research.

Throughput

As a serial technique in which samples are processed individually, CE has difficulty competing with slab gel electrophoresis where multiple samples are run in parallel. Chromatographers who are accustomed to serial analyis on automated instrumentation have been receptive to CE as an alternative technique. In conrast, CE has made few inroads in the electrophoresis community. Parallel processing of multiple samples in capillary arrays or channels etched in glass, silicon, or plastic substrates will combine the instrumental features of CE with the throughput of gel electrophoresis. Graphical presentation of data in an "electrophoresis-friendly" format in addition to numerical analysis using chromatography-based data reduction software may help CE to gain acceptance among slab gel users. These advances will certainly be important in clinical diagnostics, where sample throughput is critical in high-volume labs and gel-based serum protein analysis is a recognized diagnostic tool.

Considering the expanding CE literature and the continuing evolution of CE instrumentation, we can anticipate that many of the technical problems in application of CE to protein analysis will find creative solutions in the near future, and we are optimistic that routine application of CE in the protein laboratory will grow rapidly.

References

1. P. D. Grossman and J. C. Colburn (eds.), *Capillary Electrophoresis: Theory and Practice*, Academic Press, San Diego, 1992.
2. S. F. Y. Li, *Capillary Electrophoresis: Principles, Practice, and Applications*, J. Chromatogr. Library, Vol. 52, Elsevier Science Publisher, Amsterdam. 1992.
3. R. Weinberger, *Practical Capillary Electrophoresis*, Academic Press, Boston, 1992.
4. James P. Landers (ed.), *Handbook of Capillary Electrophoresis*, 2nd ed., CRC Press, Boca Raton,1996.
5. P. Camilleri (ed.), *Capillary Electrophoresis, Theory and Practice*, CRC Press, Boca Raton, 1993.
6. N. A. Guzman, *Capillary Electrophoresis Technology*, Marcel Dekker, New York, 1993.
7. J. Vindevogel, *Introduction to Micellar Electrokinetic Chromatography*, Hüthig, 1992.
8. S. M. Lunte and D. M. Radzik (eds.), *Pharmaceutical and Biomedical Applications of Capillary Electrophoresis*, Progress in Pharmaceuical and Biomedical Analysis, Vol. 2, Elsevier Science, Ltd. Oxford, 1996.
9. W. G. Kuhr, *Anal. Chem., 62*: 403R (1990).
10. W. G. Kuhr and C. A. Monnig, *Anal. Chem., 64:* 389R (1992).
11. C. A. Monnig and R. T. Kennedy, *Anal. Chem., 66*: 280R (1994).
12. M. V. Novotny, K. A. Cobb, and J. Liu, *Electrophoresis, 11*: 735 (1990).
13. Z. Deyl and R. Struzinsky, *J. Chromatogr., 569*: 63 (1991).
14. C. Schöneich, S. K. Kwok, G. S. Wilson, S. R. Rabel, J. F. Stobaugh,T. D. Williams, and D. G. Vander Velde, *Anal. Chem., 65*: 67R (1993).
15. C. Schöneich, A. F. R. Hühmer, S. R. Rabel, J. F. Stobaugh, S. D. S. Jois, C. K. Larive, T. J. Siahaan, T. C. Squier, D. J. Bigelow, and T. D. Williams, *Anal. Chem., 67:* 155R (1995).

263

16. É. Szökö, *Electrophoresis 18:* 74 (1997).
17. S. Hjertén, *Nucleosides and Nucleotides, 9:* 319 (1990).
18. S. Hjertén, S. Jerstedt, and A. Tiselius, *Anal. Biochem., 11:* 219 (1965).
19. D. S. Burgi and R.-L. Chien, *Anal. Chem., 63:* 2042 (1991).
20. R. Aebersold and H. D. Morrison, *J. Chromatogr., 516:* 79 (1990).
21. N. A. Guzman, J. Moschera, C. A. Bailey, K. Iqbal, and A. W. Malick, *J. Chromatogr., 598:* 123 (1992).
22. E. L. Gump and C. A. Monnig, *J. Chromatogr. A, 715:* 167 (1995).
23. D. F. Swaile and M. J. Sepaniak, *J. Liq. Chromatogr., 14:* 869 (1991).
24. T. T. Lee and E. S. Yeung, *J. Chromatogr., 595:* 319 (1992).
25. W. G. Kuhr and E. S. Yeung, *Anal. Chem., 60:* 2642 (1988).
26. S. Hjertén, K. Elenbring, F. Kilar, J.-L. Liao, A. Chen, C. J. Siebert, and M. Zhu, *J. Chromatogr., 403:* 47 (1987).
27. B. Nickerson and J. W. Jorgenson, *J. Chromatogr. 480:* 157 (1989).
28. E. T. Wise, N. Singh, and B. L. Hogan, *J. Chromatogr. A,, 746:* 109 (1996).
29. D. M. Pinto, E. A. Arriaga, D. Craig, J. Angelova, N. Sharma, H. Ahmadzadeh, and N. Dovichi, *Anal. Chem. 69:* 3015 (1997).
30. J. K. Abler, K. R. Reddy, and C. S. Lee, *J. Chromatogr. A, 759:* 139 (1997).
31. J. Cai and J. Henion, *J. Chromatogr. 703:* 667 (1995).
32. M. Yamashita and J. B. Fenn, *J. Phys. Chem., 88:* 4451 (1988).
33. R. D. Smith, C. J. Barinaga, and H.R. Udseth, *Anal. Chem., 60:* 1948 (1988).
34. E. D. Lee, W. Muck, J. D. Henion, and T.R. Covey, *J. Chromatogr. 458:* 313 (1988).
35. R. M. Caprioli, W. T. Moore, M. Martin, B. B. DaGue, K. Wilson, and S. Moring, J. Chromatogr., 480: 247 (1989).
36. M. A. Moseley, L. J. Deterding, K. B. Tomer, and J. W. Jorgenson, *J. Chromatogr.*, 480: 197 (1989).
37. T. J. Thompson, F. Foret, D. P. Kirby, P. Vouros, and B. L. Karger, presented at the 41st ASMS Conference on Mass Specrometry and Allied Topics, San Francisco, CA , 1993.
38. J. H. Wahl and R. D. Smith, *J. Cap. Elec., 1:* 62 217 (1994).
39. L. Fang, R. Zhang, E. R. Williams, and R. N. Zare, *Anal. Chem. 66:* 3696 (1994).
40. D. Figeys, I. van Oostveen, A. Ducret, and R. Aebersold, *Anal. Chem., 68:* 1822 (1996).
41. D. Figeys, A. Ducret, and R. Aebersold, *J. Chromatogr. A, 763:* 295 (1997).

42. J. H. Wahl and R. D. Smith, *J. Cap. Elec., 1:* 62 (1994).
43. J. C. Severs, S. A. Hofstadler, Z. Zhao, R. T. Senh, and R. D. Smith, *Electrophoresis, 17:* 1808 (1996).
44. H. H. Lauer and D. McManigill, *Anal. Chem., 58:* 166 (1986).
45. R. M. McCormick, *Anal. Chem., 60:* 2322 (1988).
46. J. S. Green and J. W. Jorgenson, *J. Chromatogr., 478:* 63 (1989).
47. F. A. Chen, L. Kelly, R. Palmieri, R. Biehler, and H. Schwartz, *J. Liq. Chromatogr., 15:* 1143 (1992).
48. L. Song, Q. Ou, and W. Yu, *J. Liq. Chromatogr., 17:* 1953 (1994).
49. N. Cohen and E. Grushka, *J. Chromatogr. A, 678:* 167 (1994).
50. D. Corradini, A. Rhomberg, and C. Corradini, *J. Chromatogr. A, 661:* 305 (1994).
51. D. Corradini, G. Cannarsa, E. Fabbri, and C. Corradini, *J. Chromatogr., A 709:* 127 (1995).
52. J. P. Landers, R. P. Oda, B. J. Madden, and T. C. Spelsberg, *Anal. Biochem., 205:* 115 (1992).
53. R. P. Oda, B. J. Madden, T. C. Spelsberg, and J. P. Landers, J. P., *J. Chromatogr. A, 680:* 85 (1994).
54. M. M. Bushey and J. W. Jorgenson, *J. Chromatogr., 480:* 301 (1989).
55. A. Emmer, M. Jansson, and J. Roeraade, *J. Chromatogr., 547:* 544 (1991).
56. A. Emmer and J. Roeraade, *J. Liq. Chromatogr., 17:* 3831 (1994).
57. A. Emmer, M. Jasson and J. Roerrade, *J. High Res. Chromatogr. 14:* 778 (1991).
58. E. L. Hult, A. Emmer, and J. Roeraade, *J. Chromatogr. A, 757:* 255 (1997).
59. V. Rohlicek and Z. Deyl, *J. Chromatogr., 494:* 87 (1989).
60. M. J. Gordon, K.-J. Lee, A. A. Arias, and R. N. Zare, *Anal. Chem., 63:* 69 (1991).
61. W. G. H. M. Muijselaar, C. H. M. M. de Bruijn, and F. M. Everaerts, *J. Chromatogr., 605:* 115 (1992).
62. G. Mandrup, *J. Chromatogr., 604:* 267 (1992).
63. B. Y. Gong and J. W. Ho, *Electrophoresis 18:* 732 (1997).
64. M. Taverna, A. Baillet, D. Biou, M. Schlüter, R. Werner, and D. Ferrier, *Electrophoresis, 13:* 359 (1994).
65. M. E. Legaz and M. M. Pedrosa, *J. Chromatogr. A, 719:* 159 (1996).
66. H. G. Lee and D. M. Desiderio, *J. Chromatogr. B, 691:* 67 (1997).
67. N. Guzman, J. Moschera, K. Iqbal, and A. W. Malick, *J. Chromatogr., 608 :* 197 (1992).

68. X.-H. Fang, T. Zhu, and V.-H. Sun, *J. High Res. Chromatogr., 17:* 749-(1994).
69. M. A. Strege and A. L. Lagu, *J. Liquid Chromatogr., 16:* 51 (1993).
70. M. A. Strege and A. L. Lagu, *J. Chromatogr., 630:* 337 (1993).
71. A. Cifuentes, M. A. Rodriguez, and F. G. García-Montelongo, *J. Chromatogr. A, 742:* 257 (1996).
72. Y. J. Yao and S. F. Y. Li., *J. Chromatogr. A, 663:* 97 (1994).
73. N. Cohen, and E. Grushka, *J. Cap. Elec., 1:* 112 (1994).
74. M. Morand, D. Blaas, and E. Kenndler, *J. Chromatogr. B, 691:* 192 (1997).
75. G. N. Okafo, H. C. Birrell, M. Greenaway, M. Haran, and P. Camilleri, *Anal. Biochem., 219:* 201 (1994).
76. G. N. Okafo, A. Vinther, T. Kornfelt, and P. Camilleri, *Electrophoresis, 16:* 1917 (1995).
77. J. R. Veraart, Y. Schouten, C. Gooijer, and H. Lingeman, *J. Chromatogr. A: 768:* 307 (1997).
78. A. Cifuentes, J. M. Santos, M. de Frutos, and J. C. Diez-Mesa, *J. Chromatogr. A, 652:* 161 (1993)
79. T. Wehr, *LG-GC,11:* 14 (1993).
80. K. A. Turner, *LC-GC, 9:* 350 (1991).
81. J. R. Mazzeo and I. S. Krull, *Biochromatography, 10:* 638 (1991).
82. J. E. Wiktorowicz and J. C. Colburn, *Electrophoresis, 11*: 769 (1990).
83. J. K. Towns and F. E. Regnier, *J. Chromatogr. 516:* 69 (1990).
84. A. Cifuentes, H. Poppe, J. C. Kraak, and F. B. Erim, *J. Chromatogr. B, 681:* 21 (1996).
85. D. Belder and G. Schomburg, *J. High Res. Chromatogr., 15*: 686 (1992).
86. M. Gilges, M. H. Kleesmiss, and G. Schomburg, *Anal. Chem., 66:* 2038 (1994).
87. N. Iki and E. S. Yeung, *J. Chromatogr. A, 731:* 273 (1996).
88. M. H. A. Busch, J. C. Kraak, and H. Poppe, *J. Chromatogr., A 695:* 287 (1995).
89. S. Hjerten, *J. Chromatogr., 347:* 191 (1985).
90. K. Cobb, V. Dolnik, and M. Novotny, *Anal. Chem., 62*: 2478 (1990).
91. M. Huang, J. Plocek, and M.V. Novotny, *Electrophoresis, 16:*396 (1995)
92. D. Schmalzing, C. A. Piggee, F. Foret, E. Carrilho, and B. L. Karger, *J.Chromatogr. A, 652:* 149 (1993).
93. P. Sun, A. Landman, G. E. Barker, and R. A. Hartwick, *J. Chromatogr. A, 685:* 303 (1994).

94. R. J. Xu, C. Vidal-Madjar, B. Sébille, and J. C. Diez-Masa, *J. Chromatogr. A, 730:* 289 (1996).
95. Y. Liu, R. Fu, and J. Gu, *J. Chromatogr. A, 723:* 157 (1996).
96. K. Srinivasan, C. Pohl, and N. Avdalovic, *Anal. Chem. 69:* 2798 (1997).
97. S. Hjertén and K. Kubo, *Electrophoresis, 14:* 390 (1993).
98. Y. Mechref and Z. El Rassi, *Electrophoresis, 16:* 617 (1995).
99. J. L. Liao, J. Abramson, and S. Hjertén, *J. Cap. Electrophoresis, 2:* 4 (1995).
100. M. Chiari, C. Micheletti, M. Nesi, M. Fazio, and P. G. Righetti, *Electrophoresis, 15:* 177 (1994).
101. J. K. Towns and F. E. Regnier, *Anal. Chem., 63:* 1126 (1991).
102. R. L. Chien and D. S. Burgi, *J. Chromatogr., 559:* 141 (1991)
103. P. Jandik and W. R. Jones, *J. Chromatogr., 546:* 431 (1991).
104. N. A. Guzman, M. A. Trebilock, and J. P. Advis, *J. Liq. Chromatogr., 14:* 997 (1991).
105. S. Hjertén, K. Elenbring, F. Kilár, J. L. Liao, A. J. C. Chen, C. J. Siebert, and M. Zhu, *J. Chromatogr., 403:* 47 (1987).
106. F. Foret, E. Szoko, and B. L. Karger, *Electrophoresis, 14:* 417 (1993).
107. S. Hjertén, J-L. Liao, and R. Zhang, *J. Chromatogr. A, 676:* 409 (1994).
108. R. Chien and D. S. Burgi, *Anal. Chem., 64*: 489A (1992).
109. S. Hjertén, L. Valtcheva, and Y.-M. Li, *J. Cap. Electrophoresis., 1:* 83 (1994).
110. F. M. Everaerts, J. L. Beckers, and Th.P.E.M. Verheggen, *Isotachophoresis*, J. Chromatogr. Library, Vol. 6, Elsevier Science Publisher, Amsterdam, 1976, pp. 7–23.
111. D. T. Witte, S. Nagard, and M. Larsson, *J. Chromatogr. A, 687:* 155 (1994).
112. J.-L. Liao, R. Zhang, and S. Hjertén, *J. Chromatogr. A, 676:* 421 (1994).
113. N. A. Guzman, M. A. Trebilcock, and J. P. Davis, *J. Liq. Chromatogr., 14:* 997 (1991).
114. J. H. Beattie, R. Self, and M. P. Richards, *Electrophoresis, 16:* 322 (1995).
115. L. J. Cole and R. T. Kennedy, *Electrophoresis, 16:* 549 (1995).
116. T. Hirokawa, A. Ohmori, and Y. Kiso, *J. Chromatogr. A, 634:* 101 (1993).

117. K. Ibel, R. P. May, K. Kirschner, H. Szadkowski, E. Mascher, and P. Lundahl, *Eur. J. Biochem., 190:* 311 (1990).
118. M. A. Strege and A. L. Lagu, *J. Chromatogr. A, 780:* 285 (1997).
119. J. Bao and F. E. Regnier, *J. Chromatogr,. 608*: 217 (1992).
120. D. Wu and F. E. Regnier, *Anal. Chem., 65:* 2029 (1993).
121. B. J. Harmon, D. H. Patterson, and F. E. Regnier, *Anal. Chem., 65:* 2655 (1993).
122. D. H. Patterson, B. H. Harmon, and F. E. Regnier, *J. Chromatogr. A, 732:* 119 (1996).
123. D. Wu, F. E. Regnier, and M. C. Linhares, *J. Chromatogr. B, 657:* 357 (1994).
124. D. S. Zhao and F. A. Gomez, *Chromatographia 44:* 514 (1997).
125. M. F. Regehr and F. E. Regnier, *J. Cap. Elec., 3:* 117 (1996).
126. J. Saevels, A. Van Schepdael, and J. Hoogmartens, *Electrophoresis, 17*:1222 (1996).
127. J. M. Fujima and N. D. Danielson, *J. Cap. Elec. 6:* 281 (1996).
128. P. Sun and R. A. Hartwick, *J. Chromatogr., 695:* 279 (1995).
129. Q. Xue and E. S. Yeung, *Anal. Chem. 66:* 1175 (1994).
130. Q. Xue and E. S. Yeung, *J. Chromatogr. B, 677:* 233 (1996).
131. K. Shimura and K. Kasai, *Anal. Biochem. 251:* 1 (1997).
132. Y. H. Chu, L. Z. Avila, H. A. Biebuyck, and G. M. Whitesides, *J. Med. Chem., 35:* 2915 (1992).
133. L. Z. Avila, Y. H. Chu, E. C. Blossey, and G. M. Whitesides, *J. Med. Chem., 36:* 126 (1993).
134. F. A. Gomez, J. N. Mirkovich, V. M. Dominguez, K. W. Liu, and D. M. Macias, *J.Chromatogr. A, 727:* 291 (1996).
135. F. A. Gomez, L. Z. Avila, Y. H. Chu, and G. M. Whitesides, *Anal. Chem., 66:* 1785 (1994).
136. J. C. Kraak, S. Busch, and H. Poppe, *J. Chromatogr., 608:* 257 (1992).
137. M. H. A. Busch, H. F. M. Boelens, J. C. Kraak, H. Poppe, A. A. P. Meekel, and M. Resmini, *J. Chromatogr. A, 744:* 195 (1996).
138. M. H. A. Busch, H. F. M. Boelens, J. C. Kraak, and H. Poppe, *J. Chromatogr. A, 775:* 313 (1997).
139. S. Honda, A. Taga, K. Suzuki, S. Suzuki, and K. Kakehi, *J. Chromatogr., 597:* 377 (1992).
140. R. Kuhn, R. Frei, and M. Christen, *Anal. Biochem., 218:* 131 (1994).
141. N. H. Heegaard and J. Brimnes, *Electrophoresis 17:* 1916 (1996).
142. N. H. H. Heegaard, D. T. Olsen, and K.-L. P. Larsen, *J. Chromatogr. A, 744:* 285 (1996).

143. Y. H. Chu, J. L. Watson, A. Stassinopoulos, and C. T. Walsh, *Biochemistry, 33:* 10616 (1994).

144. R. G. Nielsen, E. C. Rickard, P. F. Santa, D. A. Sharknas, and G. S. Sittampalam, *J. Chromatogr., 539:* 177 (1991).

145. O. W. Reif, R. Lausch, T. Scheper, and R. Freitag, *Anal. Chem., 66:* 4027 (1994).

146. J. Pédron, R. Maldiney, M. Brault, and E. Miginiac, *J. Chromatogr. A, 723:* 381 (1996).

147. N. M. Schultz and R. Kennedy, *Anal. Chem. 67:* 924 (1995).

148. L. Tao and R. Kennedy, *Electrophoresis, 18:* 112 (1997).

149. L. Tao and R. Kennedy, *Anal. Chem. 68:* 3899 (1996).

150. N. Chiem and D. J. Harrison, *Anal. Chem. 69:* 373 (1997).

151. K. Nadeau, S. G. Nadler, M. Saulnier, M. A. Tepper, and C. T. Walsh, *Biochemistry, 33:* 2561 (1994).

152. J. Liu, K. J. Volk, M. S. Lee, E. H. Kerns, and I. E. Rosenberg, *J. Chromatogr. A, 680:* 395 (1994).

153. P. Sun, A. Hoops, and R. A. Hartwick, *J. Chromatogr. B, 661:* 335 (1994).

154. V. T. Chadwick, A. C. Cater, A. C., and J. J. Wheeler, *J. Liq. Chromatogr., 16:* 1903 (1993).

155. J. C. Olivier, M. Taverna, C. Vauthier, P. Couvreur, and D. Baylocq-Ferrier, *Electrophoresis, 15:* 234 (1994).

156. F.-T. A. Chen and J. C. Sternberg, Electrophoresis 15 (1994) 13.

157. V.M. Okun and U. Bilitewski, *Electrophoresis 17:* 1627 (1996).

158. T. J. Pritchett, R. A. Evangelista, and F.-T. A. Chen, *J. Cap. Elec., 2:* 145 (1995).

159. M. J. Schmerr, K. R. Goodwin, R. C. Cutlip, and A. L. Jenny, *J. Chromatogr. B, 681:* 29 (1996).

160. H. Frokiaer, M. Mortensen, H. Sorensen, and S. Sorensen, *J. Liq. Chrom. & Rel. Technol. 19:* 57 (1996).

161. K. Haupt, F. Roy, and M. A. Vijayalakshmi, *Anal. Biochem., 234:* 149 (1996).

162. M. Novotny, H. Soini, and M. Stefansson, *Anal. Chem. 66:* 646A (1994).

163. S. G. Allenmark and S. Andersson, *J. Chromatogr. A, 666:* 167 (1994).

164. H. Nishi and S. Terabe, *J. Chromatogr. A, 694:* 245 (1995).

165. D. K. Lloyd, S. Li, and P. Ryan, *J. Chromatogr. A, 694:* 285 (1995).

166. S. Birnbaum and S. Nilsson, *Anal. Chem., 64:* 2872 (1992)

167. G. E. Barker, P. Russo, and R. A. Hartwick, *Anal. Chem., 64:* 3024 (1992).

168. P. Sun, N. Wu, G. Barker, and R. A. Hartwick, *J. Chromatogr., 648:* 475 (1993).

169. P. Sun, G. Barker, and R. A. Hartwick, *J. Chromatogr. A, 652:* 247 (1993).

170. R. Vespalec, V. Sustácek, and P. Bocek, *J. Chromatogr., 638:* 255 (1993).

171. T. Arai, N. Nimura, and T. Kinoshita, *Biomed. Chromatogr, 9:* 68 (1995).

172. T. Arai, M. Ichinose, H. Kuroda, N. Nimura, and T. Kinoshita, *Anal. Biochem., 217:* 7 (1994).

173. S. Li and D. K. Lloyd, *Anal. Chem., 65:* 3684 (1993).

174. Y. Ishihama, Y. Oda, N. Asakawa, Y. Yoshida, and T. Sato, *J. Chromatogr. A, 666:* 193 (1994).

175. Y. Tanaka, N. Matsubara, and S.Terabe, *Electrophoresis, 15:* 848 (1994).

176. S. Busch, J. C. Kraak, and H. Poppe, *J. Chromatogr., 635:* 119 (1993).

177. D. Wistuba, H. Diebold, and V. Schurig, *J. Microcol. Sep., 7:* 17 (1995).

178. L. Valtcheva, J. Mohammad, G. Pettersson, and S. Hjertén, *J. Chromatogr., 638:* 263 (1993).

179. Y. Tanaka and S. Terabe, *J. Chromatogr. A, 694:* 277 (1995).

180. Y. Tanaka and S. Terabe, *Chromatographia 44:* 119 (1997).

181. R. S. Rush, A. Cohen, and B. L. Karger, *Anal. Chem., 63:* 1346 (1991).

182. M. A. Strege and A. L. Lagu, *J. Chromatogr. A, 652:* 179 (1993).

183. M. A. Strege and A. L. Lagu, *American Lab. 26:* 48C (1994)

184. F. Kilár and S. Hjertén, *J. Chromatogr., 638:* 269 (1993).

185. V. J. Hilser, G. D. Worosila, and E. Freire, *Anal. Biochem., 208:* 125 (1993).

186. V. J. Hilser and E. Freire, *Anal. Biochem., 224:* 465 (1995).

187. R. T. Bishop, V. E. Turula, and J. A. de Haseth, *Anal. Chem. 68:* 4006 (1996).

188. M. Kats, P. C. Richberg, and D. E. Hughes, *Anal. Chem., 67:* 2943 (1995).

189. M. Kats, P. C. Richberg, and D. E. Hughes, *Anal. Chem., 69:* 338 (1997).

190. M. Kats, P. C. Richberg, and D. E. Hughes, *J. Chromatogr. A, 766:* 205 (1997).

191. Z. H. Fan, P. K. Jensen, C. S. Lee, and J. King, *J. Chromatogr. A, 769:* 315 (1997).
192. R. Rodriguez-Diaz, M. Zhu, V. Levi, R. Jimenez, and T. Wehr, presented at the 7th Symposium on Capillary Electrophoresis, Würzburg, Germany, 1995.
193. N. M. Kinghorn, C. S. Norris, G. R. Paterson and D. E. Otter, *J. Chromatogr. A, 700:* 111 (1995).
194. G. R. Paterson, J. P. Hill, and D. E. Otter, *J. Chromatogr. A, 700*: 105 (1995).
195. N. M. Kinghorn, G. R. Paterson, and D. E. Otter, *J. Chromatogr. A, 723:* 371 (1997).
196. I. Recio, E. Molina, M. Ramos, and M. de Frutos, *Electrophoresis, 16:* 654 (1995).
197. R. Jimenez-Flores and A. Ulibarri, *J. Dairy Science, 78* (supplement 1): 145 (1995).
198. K. R. Kristiansen, J. Otte, M. Zakora, and K. B. Qvist, *Milchwissenscaft, 49:* 683 (1994).
199. T. M. I. E. Christensen, K. R. Kristiansen, and J. S. Madsen, *J. Dairy Res., 56:* (5) 823-828 (1989).
200. M. Kanning, M. Castella, and C. Olieman, *LC-GC, 6:* 701 (1993).
201. I. Recio, M.-L. Pérez-Rodriguez, M. Ramos, and L. Amigo, *J. Chromatogr. A, 768:* 47 (1997).
202. B. I. Recio and C. Olieman, *Electrophoresis, 17:* 1228 (1996).
203. F-T. A. Chen and A. Tusak, *J. Chromatogr. A, 685:* 331 (1994).
204. M. P. Richards and J. H. Beattie, *J. Cap. Elect., 1:* 196 (1994).
205. R. S. Rush, A. S. Cohen, and B.L. Karger, *Anal. Chem., 63:* 1346 (1991).
206. T. T. Lee and E. S. Yeung, *Anal.Chem., 64:* 3045 (1992).
207. H. Kajiwara, *J. Chromatogr., 559:* 345 (1991).
208. J. H. Beattie, M. P. Richards, and R. Self, *J. Chromatogr., 632:* 127 (1993).
209. G.-Q. Liu, W. Wang, and X.-Q. Shan, *J. Chromatogr. B, 653:* 41 (1994).
210. V. Virtanen, G. Bordin, and A.-R. Rodriguez, *J. Chromatogr. A, 734:* 391 (1996)
211. M. P. Richards and P. J. Aagaard, *J. Cap. Elec., 1:* 90 (1994).
212. M. P. Richards and J. H. Beattie, *J. Chromatogr. B, 669:* 27 (1995).
213. M. P. Richards, *J. Chromatogr. B, 657:* 345 (1994).
214. M. P. Richards, G. K. Andrews, D. R. Winge, and J. H. Beattie, *J. Chromatogr. B, 675:* 327 (1996).

215. T. Minami, H. Matsubara, M. Ohigashi, K. Kubo, N. Okabe, and Y. Okazaki, *Electrophoresis, 17:* 1602 (1996).

216. T. Minami, H. Matsubara, M. Ohigashi, N. Otaki, M. Kimura, K. Kubo, N. Okabe, and Y. Okazaki, *J. Chromatogr. B, 685:* 353 (1996).

217. E. Torres, A. Cid, P. Fidalgo, and J. Abalde, *J. Chromatogr. A, 775:* 339 (1997).

218. J. H. Beattie and M. P. Richards, *J. Chromatogr., 664:* 129 (1994).

219. Z. Zhao, A. Malik, M. L. Lee, and G. D. Watt, *Anal. Biochem., 218:* 47 (1994).

220. F. S. Markland, S. Morris, J. R. Deschamps, and K. B. Ward, *J. Liq. Chromatogr., 16:* 2189 (1993).

221. J. Y. Cai and Z. El Rassi, *J. Liq. Chromatogr., 16:* (1993) 2007 (1993).

222. F. Kilár and S. Hjertén, *J. Chromatogr., 480:* 351 (1989).

223 K. Yim, *J. Chromatogr., 559:* 401(1991).

224. J. M. Thorne, W. K. Goetzinger, A. B. Chen, K. G. Moorhouse, and B. L. Karger, *J. Chromatogr. A, 744* : 155 (1996).

225. A. D. Tran, S. Park, and P. J. Lisi, *J. Chromatogr., 542:* 459 (1991).

226. E. Watson and F. Yao, *Anal. Biochem., 210:* 389 (1993).

227. H. P. Bietlot and M. Girard, *J. Chromatogr. A, 759:* 177 (1997).

228. D. E. Morbeck, B. J. Madden, and D. McCormick, *J. Chromatogr. A, 680:* 217 (1994).

229. P. M. Rudd, H. C. Joao, E. Coghill, P. Fiten, M. R. Saunders, G. Opdenakker, and R.A. Dwek, *Biochemistry, 33:* 17 (1994).

230. K. Yim, J. Abrams, and A. Hsu, *J. Chromatogr. A, 716:* 401 (1995).

231. Y. Chen, *J. Chromatogr. A, 768:* 39 (1997).

232. J. A. Bietz, "Fractionation of wheat gluten proteins by capillary electrophoresis," in *Gluten Proteins*, Association of Cereal Research: Detmold, Germany, 1993, pp. 404–413.

233. J. A. Bietz and E. Schmalzried, *Lebensm.-Wiss. u.-Technol., 28*: 174 (1995).

234. W. E. Werner, J. E. Wiktorowicz, and D. D. Kasarda, *Cereal Chem., 71:* 397 (1994).

235. G. L. Lookhart and S. R. Bean, *Cereal Chem., 72:* 42 (1995).

236. G. L. Lookhart and S. Bean, *Cereal Chem., 72:* 527 (1995).

237. G. L. Lookhart and S. R. Bean, *Cereal Chem., 72:* 312 (1995).

238. G. L. Lookhart and S. R. Bean, *Cereal Chem., 73:* 81(1996).

239. G. L. Lookhart, S. R. Bean, R. Graybosch, O. K. Chung, B. Morena-Sevilla, and S. Baenziger, *Cereal Chem., 73:* 547 (1996).

240. I. Shomer, G. Lookhart, R. Salomon, R. Vasiliver, and S. Bean, *J. Cereal Sci., 22:* 237 (1995).

241. T. M. Wong, C. M. Carey and S. H. C. Lin, *J. Chromatogr. A, 680:* 413 (1994).
242. Z. Deyl, *J. Chromatogr. 488:* 161 (1989).
243. Z. Deyl and V. Rohlicek, *J. Chromatogr. 480:* 371 (1989).
244. Z. Deyl and I. Miksík, *J. Chromatogr. 698:* 369 (1995).
245. Z. Deyl, V. Rohlicek, and R. Struzinsky, *J. Liq. Chromatogr. 12: 3515* (1989).
246. J. Novotná, Z. Deyl, and I. Miksík, *J. Chromatogr. B, 681:* 77 (1996).
247. Z. Deyl, J. Novotná, I. Miksík, D. Jelínková, M. Uhrová, and M. Suchánek, *J. Chromatogr. B, 689:* 181 (1997).
248. I. Miksík, J. Novotná, M. Uhrová, D. Jelínková, and Z. Deyl, *J. Chromatogr. A, 772:* 213 (1997).
249. T. M. McNerney, S. K. Watson, J.-H. Sim, and R. L. Bridenbaugh, *J. Chromatogr. A, 744:* 223 (1996).
250. R. Malsch, T. Mrotzek, G. Huhle, and J. Harenberg, *J. Chromatogr. A, 744:* 215 (1996).
251. S. Kundu, C. Fenters, M. Lopez, B. Calfin, M. Winkler, and W. G. Robey, *J. Cap. Elec., 6:* 301 (1996).
252. E. Jellum, J. Cap. Elec. 1: 97 (1994).
253. M. A. Jenkins and M. D. Guerin, *J. Chromatogr. B, 682:* 23 (1996).
254. S. Hjertén, *Chromatogr. Rev., 9:*122 (1967).
255. J. W. Jorgenson and K. D. Lukacs, *Science, 222:* 266 (1983).
256. F.-T. A. Chen, *J. Chromatogr. 559:* 445 (1991).
257. F.-T.A. Chen, C.-M. Liu, Y.-Z. Hsieh, and J. C. Sternberg, *Clin. Chem., 37:* 14 (1991).
258. E. Jellum, H. Dollekamp, and C. Blessum, *J. Chromatogr. B, 683:* 55 (1996).
259. E. Jellum, H. Dollekamp, A. Brunsvig, and R. Gislefoss, *J. Chromatogr. B, 689:* 155 (1997).
260. J. W. Kim, J. H. Park, J. W. Park, H. J. Doh, G. S. Heo, and K.-J. Lee, *Clin. Chem., 39:* 689 (1993).
261. R. Lehmann, H. Liebich, G. Grübler, and W. Voelter, *Electrophoresis 16 :* 998 (1995).
262. O. W. Reif, R. Lausch, and R. Freitag, *American Laboratory, May:* 33 (1994).
263. M. A. Jenkins, T. D. O'Leary, and M. D. Guerin, *J. Chromatogr. B, 662:* 108 (1994).
264. M. A. Jenkins, E. Kulinskaya, H. D. Martin, and M. D. Guerin, *J. Chromatogr. B, 672:* 241 (1995).
265. V. Dolnik, *J. Chromatogr. A, 709:* 99 (1995).

266. R. Clark, J. A. Katzmann, E. Wiegert, C. Namyst-Goldberg, L. Sanders, R. P.Oda, R. A. Kyle, and J. P. Landers, *J. Chromatogr. A., 744:* 205 (1996).

267. R. Lehmann, M. Koch, W. Voelter, H. U. Häring, and H. M. Liebich, *Chromatographia, 45:* 390 (1997).

268. K. Andrieux, J. C. Olivier, M. Taverna, C. Vauthier, P. Couvreur, and D. Ferrier, *J. Liq. Chrom. & Rel. Technol., 19:* 3333 (1996).

269. Z. K. Shihabi, *Electrophoresis 17:* 1607 (1996).

270. G. L. Klein and C. R. Joliff, in J. P. Landers (ed.), *Handbook of Capillary Electrophoresis*, CRC Press, Boca Raton, FL, 1993, p. 452.

271. P. Ferranti, A. Malorni, P. Pucci, S. Fanali, A. Nardi, and L. Ossicini, *Anal. Biochem. 194:* 1 (1991).

272. C.-N. Ong, L. S. Liau, and H. Y. Ong, *J. Chromatogr. 576:* 346 (1992).

273. M. Zhu, R. Rodriguez, T. Wehr, and C. Siebert, *J. Chromatogr. 608:* 225 (1992).

274. M. Zhu, T. Wehr, V. Levi, R. Rodriguez, K. Shiffer, and Z. A. Cao, *J. Chromatogr. A, 652:* 119 (1993).

275. D. Josíc, A. Böttcher, and G. Schmitz, *Chromatographia, 30:* 703 (1990).

276. G. Schmitz and C. Möllers, *Electrophoresis,15:* 31 (1994).

277. T. Tadey and W. C. Purdy, *J. Chromatogr., 583:* 111 (1992).

278. T. Tadey and W. C. Purdy, *J. Chromatogr. A, 652:* 131 (1993).

279. I. D. Cruzado, A. Z. Hu, and R. D. Macfarlane, *J. Capillary Electrophor., 3:* 25 (1996)

280. I. D. Cruzado, S. Song, S. F. Crouse, B. C. O'Brien, and R. D. Macfarlane, *Anal. Biochem., 243:* 100 (1996).

281. Z. K. Shihabi, *J. Chromatogr. B, 669:* 53 (1995).

282. P. G. Righetti, *Isoelectric Focusing: Theory, Methodology and Applications*, Elsevier Biomedical Press, Amsterdam, 1989.

283. A. Kolin, *Electrophoresis, 4:* 1(1983).

284. T. J. Pritchett, *Electrophoresis , 17:* 1195(1996).

285. S.Hjertén, "Isoelectric Focusing in Capillaries," in *Capillary Electrophoresis: Theory and Practice*, P. D. Grossman and J. C. Colburn, (ed.), Academic Press, San Diego, CA, 1992, chap. 7.

286. X. Liu, Z. Sosic, and I. S. Krull, *J. Chromatogr. A, 735:* 165 (1996).

287. J. R. Mazzeo and I. S. Krull, *J. Microcol. Sep., 4:* 29 (1992).

288. S.-M. Chen and J. E. Wiktorowicz, *Anal. Biochem., 206:* 84 (1992).

289. F. Kilar and S. Hjertén, *Electrophoresis, 10:* 23 (1989).

290. R. A. Mosher and W. Thormann, *The Dynamics of Electrophoresis*, VCH, Weinheim, 1992, chap. 7.

291. S. Hjertén and M. Zhu, *J. Chromatogr., 347:* 265 (1985).
292. S. Hjertén, J.-L. Liao, and K. Yao, *J. Chromatogr., 387:* 127 (1987).
293. S. Hjertén, and M. Zhu, *J. Chromatogr., 346:* 265 (1985).
294. T. L. Huang, P. C. H. Shieh, and N. Cooke, *Chromatographia, 39:* 543(1994).
295. R. Rodriguez and C. Siebert, poster presentation 6th International Symposium on Capillary Electrophoresis, San Diego, CA, 1994.
296. W. Thormann, J. Caslavska, S. Molteni, and J. Chmelik, *J. Chromatogr., 589:* 321 (1992).
297. J. R. Mazzeo, and I. S. Krull, *Anal. Chem., 63:* 2852 (1991).
298. M. Zhu, R. Rodriguez, and T. Wehr, *J. Chromatogr., 559:* 479 (1991).
299. W. Thormann, J. Caslavska, S. Molteni, and J. Chmelik, *J. Chromatogr. 589:* 321 (1992).
300. S. Molteni and W. Thormann, *J. Chromatogr. 638:* 187 (1993).
301. J. R. Mazzeo and I. S. Krull, *J. Chromatogr. 606:* 291 (1992).
302. C. Schwer, *Electrophoresis, 16:* 2121 (1995).
303. X-W. Yao and F. E. Regnier, *J. Chromatogr., 632:* 185 (1993).
304. X-W. Yao, D. Wu, and F. E. Regnier, *J. Chromatogr., 636:* 21 (1993).
305. M. Zhu, R. Rodriguez, D. Hansen, and T. Wehr, *J. Chromatogr., 516:* 123 (1990).
306. P. G. Righetti, G. Tudor, and K. Ek, *J. Chromatogr., 220:* 115 (1981).
307. R. Rodríguez-Díaz, M. Zhu, and T. Wehr, *J. Chromatogr. A, 772:* 145 (1997).
308. A. B. Chen, C. A. Rickel, A. Flanigan, G. Hunt, and K. G. Moorhouse, *J. Chromatogr. A, 744:* 279 (1996).
309. K. Shimura and B. L. Karger, *Anal. Chem., 66:* 9 (1994).
310. N. Wu, P. Sun, J. H. Aiken, T. Wang, C. W. Huie, and R. Hartwick, *J. Liq. Chromatogr., 16:* 2293 (1993).
311. J. Wu and J. Pawliszyn, *J. Chromatogr., 608:* 121 (1992).
312. J. Wu and J. Pawliszyn, *Electrophoresis, 14:* 469 (1993).
313. J. Wu and J. Pawliszyn, *Electrophoresis, 16:* 670 (1995).
314. J. Wu and J. Pawliszyn, *J. Chromatogr. B, 657:* 327 (1994).
315. L. Vonguyen, J. Wu, and J. Pawliszyn, *J. Chromatogr. B, 657:*333 (1994).
316. J. Wu and J. Pawliszyn, *Anal. Chem., 66:* 867 (1994).
317. J. Wu and J. Pawliszyn, *Anal. Chem., 64:* 224 (1994).
318. J. Wu and J. Pawliszyn, *Anal. Chem., 64:* 219 (1992).
319. J. Wu and J. Pawliszyn, *J. Liq. Chromatogr., 16:* 3675 (1993).
320. S. J. Lillard and E. S. Yeung, *J. Chromatogr. B, 687:* 363 (1996).

321. F. Foret, O. Muller, J. Thorne, W. Gotzinger, and B. L. Karger, *J. Chromatogr. A, 716*: 157 (1995).
322. R. Rodríguez-Díaz, M. Zhu, V. Levi, and T. Wehr, presented at the 7th Symposium on Capillary Electrophoresis, Würzburg, Germany (1995).
323. J.-L. Liao and R. Zhang, *J. Chromatogr. A, 684:* 143 (1994).
324. N. J. Clarke, A. J. Tomlinson, G. Schomburg, and S. Naylor, *Anal. Chem., 69:* 2786 (1997).
325. J. Wu and J. Pawliszyn, *Anal. Chem., 67:* 2010 (1995).
326. W. Thormann, A. Tsai, J. P. Michaud, R. A. Mosher, and M. Bier, *J. Chromatogr. 389:* 75 (1987).
327. T. Rabilloud, *Electrophoresis 17:* 813 (1997).
328. M. Minarik, F. Groiss, B. Gas, D. Blaas, and E. Kenndler, *J. Chromatogr. A, 738:* 123 (1996).
329. K. Slais and Z. Friedl, *J. Chromatogr. A, 661:* 249 (1994).
330. N. Y. Nguyen and A. Chrambach, *Anal. Biochem., 79:* 462 (1977).
331. R. Grimm, *J. Cap. Elec. 002:* 111 (1995).
332. Yim, K. W., *J. Chromatogr., 559:* 401 (1991).
333. T. Wehr, M. Zhu, R. Rodriguez, D. Burke, and K. Duncan, *American Biotech. Lab.*, September 1990.
334. J. W. Drysdale, P. G. Righetti, and H. F. Bunn, *Biochem. Biophys. Acta, 229:* 42 (1971).
335. P. G. Righetti and C. Gelfi, *J. Cap. Elec., 1:* 27 (1994).
336. S. Molteni, H. Frischknecht, and W. Thormann, *Electrophoresis, 15*: 22 (1994).
337. J. M. Hempe and R. D. Craver, *Clin. Chem., 40:* 2288 (1994).
338. S. B. Harper, W. J. Hurst, and C. M. Lang, *J. Chromatogr. B, 657:* 339 (1994).
339. M. Conti, C. Gelfi, A. Bianchi Bosisio, and P. G. Righetti, *Electrophoresis, 17:* 1590 (1996).
340. P. G. Righetti, M. Conti, and C. Gelfi, *J. Chromatogr., A 767:* 255 (1997).
341. S. Kundu and C. Fenters, *J. Cap. Elec., 2:* 6 (1995).
342. O-S. Reif and R. Freitag, *J. Chromatogr. A, 680*: 383 (1994).
343. G. Hunt, K. G. Moorhouse, and A. B. Chen, *J. Chromatogr. A, 744:* 295 (1996).
344. J. M. Thorne, W. K. Goetzinger, A. B. Chen, K. G. Moorhouse, and B. L. Karger, *J. Chromatogr. A, 744:* 155 (1996).
345. J. Kubach and R. Grimm, *J. Chromatogr. A, 737:* 281 (1996).

346. K. G. Moorhouse, C. A. Eusebio, G. Hunt, and A. B. Chen, *J. Chromatogr. A, 717:* 61 (1995).
347. M. P. Richards and T. L. Huang, *J. Chromatogr. B, 690:* 43 (1997).
348. J. Wu and J. Pawliszyn, *J. Chromatogr. A, 652:* 295 (1993).
349. T. L. Huang and M. P. Richards, *J. Chromatogr. A, 757:* 247 (1997).
350. J. R. Mazzeo, J. A. Martineau, and I. S. Krull, *Anal. Biochem., 208:* 323 (1993).
351. J. Caslavska, S. Molteni, J. Chmelik, K. Slais, F. Matulik, and W. Thormann, *J. Chromatogr. A, 680:* 549 (1994).
352. D. Wu and F. Regnier, *J. Chromatogr., 608:* 349 (1992).
353. K. Tsuji, *J. Chromatogr. A, 652:* 139 (1993).
354. A. Guttman, *Electrophoresis, 16:* 611 (1995).
355. K. Tsuji, *J. Chromatography B, 662*: 291 (1994).
356. A. S. Cohen and B. L. Karger, *J. Chromatogr., 397:* 409 (1987).
357. S. Hjertén, *Electrophoresis '83*, H. Hirai, (ed.), Walter de Gruyter & Co., New York, 1994, pp. 71–79.
358. M. Zhu, D. L. Hansen, S. Burd, and F. Gannon, *J. Chromatogr., 480*: 311 (1989).
359. S. Hjerten, T. Srichaiyo, and A. Palm, *Biomed. Chromatogr., 8:* 73 (1994).
360. M. Zhu, V. Levi, and T. Wehr, *Am. Biotech. Lab., 11:* 26 (1993).
361. C. Tanford, *The Hydrophobic Effect: Formation of Micelles and Biological Membranes*, 2nd ed., Wiley, New York, 1980, pp. 159–164.
362. M. R. Karim, S. Shinagawa, and T. Takagi, *Electrophoresis, 14:* 1141 (1994).
363. K. Sasa and K. Taked, *J. Colloid Interface Sci., 147:* 516 (1993).
364. K. Benedek and S. Thiede, *J. Chromatogr. A, 676:* 209 (1994).
365. A. Guttman, J. A. Nolan, and N. Cooke, *J. Chromatogr., 632*: 171 (1993).
366. M. Nakatani, A. Shibukawa, and T. Nakagawa, *Biol. Pharm. Bull. 16:* 1185 (1993).
367. M. Nakatani, A. Shibukawa, and T. Nakagawa, *Anal. Sci., 10:* 1 (1994).
368. M. Nakatani, A. Shibukawa, and T. Nakagawa, *J. Chromatogr. A, 672:* 213 (1994).
369. W. E. Werner, D. M. Demorest, J. Stevens, and J. E. Wiktorowicz, *Anal. Biochem., 212:* 253 (1993).
370. K. Ganzler, K. S. Greve, A. S. Cohen, B.L. Karger, A. Guttman, and N. C. Cooke, *Anal. Chem., 64:* 2665 (1992).

371. A. Guttman, P. Shieh, D. Hoang, J. Horvath, and N. Cooke, *Electrophoresis, 15:* 221 (1994).

372. K. Tsuji, *J. Chromatography A, 661:* 257 (1994).

373. E. Simo-Alfonso, M. Conti, C. Gelfi, and P. G. Righetti, *J. Chromatogr. A, 689:* 85 (1995).

374. A. Guttman, J. Horvath, and N. Cooke, *Anal. Chem., 65:* 199 (1993).

375. M. R. Karim, J-C. Janson, and T. Takagi, *Electrophoresis, 15:* 1531 (1994).

376. P. G. De Gennes, *Scaling Concepts in Polymer Chemistry*, Cornell University Press, Ithaca, NY, 1979.

377. A. E Barron, H. W. Blanch, and D. S. Soane, *Electrophoresis, 15:* 597 (1994).

378. A. E. Barron, D. S. Soane, and H. W. Blanch, *J. Chromatogr. A, 652:* 3 (1993).

379. W. E. Werner, D. M. Demorest, and J. E. Wiktorowicz, *Electrophoresis, 14:* 759 (1993).

380. A. Guttman, P. Shieh, J. Lindahl, and N. Cooke, *J. Chromatogr. A, 676:* 227 (1994).

381. W. E. Werner, *Cereal Chem, 72:* 248 (1995).

382. A. Cifuentes, M. de Frutos, and J. C. Diez-Masa, *J. Dairy Sci,. 76:* 1870 (1993).

383. S. Kundu, C. Fenters, M. Lopez, A. Varma, J. Brackett, S. Kuemmerle, and J. C. Hunt, *J. Cap. Elect. 4:* 7 (1997).

384. P. C. H. Shieh, D. Hoang, A. Guttman, and N. Cooke, *J. Chromatogr. A, 676:* 219 (1994).

385. R. Lausch, T. Scheper, O-W. Reif, J. Schosser, J. Fleischer and R. Freitag, *J. Chromatogr. A, 654:* 190 (1993).

386. A. Widhalm, C. Schwer, D. Blaas and E. Kenndler, *J. Chromatogr., 549:* 446 (1991).

Index